HPLC of macromolecules

a practical approach

Edited by
R W A Oliver

Biological Materials Analysis Research Unit, Department of Biological Sciences, University of Salford, Salford M5 4WT, UK

IRL PRESS
—at—
OXFORD UNIVERSITY PRESS
Oxford New York Tokyo

Oxford University Press
Walton Street, Oxford OX2 6DP

First published 1989
Reprinted 1989, 1991

British Library Cataloguing in Publication Data
HPLC of macromolecules.
  1. Organisms. Polymers. High performance liquid chromatography
  I. Oliver, R.W.A. (Ronald William Alfred), *1932–* II. Series
  574.19'24

Library of Congress Cataloging in Publication Data
HPLC of macromolecules.
  (Practical approach series)
  Companion volume to: Hplc of small molecules.
  Includes bibliographies and index.
  1. High performance liquid chromatography.
  2. Biopolymers—Analysis. 3. Macromolecules—Analysis.
  I. Oliver, R.W.A. II. Hplc of small molecules. III. Series.
  [DNLM: 1. Chromatography, High Pressure Liquid—
  methods. 2. Macromolecular Systems.
  3. Molecular Biology—methods. QD272.C447 H872].
  QP519.9.H53H69 1988   574.19'24   88-8238

ISBN 0 19 963020 8 (hardbound)
ISBN 0 19 963021 6 (softbound)

Previously announced as:

ISBN 1 85221 067 2 (hardbound)
ISBN 1 85221 066 4 (softbound)

Printed by Information Press Ltd, Oxford, England.

TITLES PUBLISHED IN
THE
PRACTICAL APPROACH
SERIES

Series editors:
Dr D Rickwood
Department of Biology, University of Essex
Wivenhoe Park, Colchester, Essex CO4 3SQ, UK
Dr B D Hames
Department of Biochemistry and Molecular Biology, University of Leeds
Leeds LS2 9JT

Affinity chromatography
Animal cell culture
Antibodies I & II
Biochemical toxicology
Biological membranes
Carbohydrate analysis
Cell growth and division
Centrifugation (2nd Edition)
Computers in microbiology
DNA cloning I, II & III
Drosophila
Electron microscopy
in molecular biology
Fermentation
Gel electrophoresis
of nucleic acids (2nd edition)
Gel electrophoresis of proteins
Genome analysis
HPLC of small molecules
HPLC of macromolecules
Human cytogenetics
Human genetic diseases
Immobilised cells and enzymes
Iodinated density gradient media
Light microscopy in biology
Liposomes
Lymphocytes
Lymphokines and interferons
Mammalian development
Medical bacteriology
Medical mycology

Microcomputers in biology
Microcomputers in physiology
Mitochondria
Mutagenicity testing
Neurochemistry
Nucleic acid and
protein sequence analysis
Nucleic acid hybridisation
Nucleic acids sequencing
Oligonucleotide synthesis
Photosynthesis:
energy transduction
Plant cell culture
Plant molecular biology
Plasmids
Prostaglandins
and related substances
Protein function
Protein sequencing
Protein structure
Proteolytic enzymes
Ribosomes and protein synthesis
Solid phase peptide synthesis
Spectrophotometry
and spectrofluorimetry
Steroid hormones
Teratocarcinomas
and embryonic stem cells
Transcription and translation
Virology
Yeast

Preface

This text is a companion volume to *HPLC of Small Molecules: A Practical Approach*, edited by Dr C.K.Lim. The aim of the present text is to present in a concise, experimentally explicit manner, examples of high-performance liquid chromatographic separations of large biological molecules (molecules having a molecular mass greater than 1000). Because of the differences in size of the molecules under consideration, chapters are included on the application of the techniques of size-exclusion HPLC and also of high-performance affinity chromatography in addition to the more common HPLC techniques presented in its sister volume. There are also chapters on HPLC column support materials and on HPLC instrumentation since these subjects were not included in the previous volume.

It is hoped that this text will encourage more workers to attempt to purify and analyse intact large biological molecules by HPLC rather than degrading these and then purifying the resultant small fragment molecules. It is advisable for the researcher in biological materials to gain experience in HPLC techniques as applied to macromolecules because once the initial analytical separations have been achieved these provide the basis for subsequent preparative scale separation. This means that the recent developments in recombinant DNA technology which have led to the ready production of mixtures of polypeptides and proteins in large quantities, many of which are of considerable therapeutic and industrial importance, may be exploited since these mixtures may then be purified. It is probably true to state that most reported analytical and small scale purification of such mixtures contain at least one HPLC step.

Reversed-phase (RP)-HPLC is probably still the most widely used mode of HPLC separation for such mixtures in spite of the fact that significant losses of protein may occur. Thus a wide-pore C_{18}-column has been used to purify calcitonin, a 32-amino acid peptide hormone, which had been produced in *Escherichia coli* as a fusion product of chloramphenicol acetyltransferase to calcitonin−glycine. A cation-exchange HPLC method was used to separate the polypeptide hormone urogastrone produced in *E.coli* by recombinant DNA methods. Size-exclusion HPLC has been employed to separate the large hepatitis B immunoglobulin aggregates whilst affinity HPLC has been used to purify so many hundreds of enzymes that it would be invidious to pick out any one for special mention here. For the separation of very complex mixtures obviously either specific sample pre-treatment, e.g. solvent extraction, followed by a single HPLC technique may be employed and examples of this are to be found in this text. Alternatively, combinations of the various modes of HPLC can and have been employed, for example in the purification of membrane proteins.

The range of biological macromolecules covered in the chapters on specific applications of HPLC includes polypeptides and proteins, enzymes, oligonucleotides and glycopeptides. The presentation and discussion of the HPLC separation of such a wide range of materials could only have been done by enlisting the experience of a number of researchers, both academic and industrial, who are expert in these various fields. The editor would therefore like to thank all of the invited authors for giving their time and energy to write their chapters in the first place and for their patience, understanding and co-operation which has led to its final production in the style of the Practical Approach series. Finally the editor would also like to thank all of the Instrument Companies which responded so promptly to his enquiries and especially to Messrs. Kontron for providing excellent photographs of parts of equipment not normally seen!

R.W.A.Oliver

Contributors

Y.D.Clonis
Laboratory of Biochemistry, Department of Chemistry, University of Patras, GR-26110-Patras, Greece

P.H.Corran
National Institute for Biological Standards and Control, Blanche Lane, South Mimms, Potters Bar, Herts EN6 3QG, UK

A.Fleiss
Abteilung Biophysikalische Chemie, Zentrum Biochemie, Medizinische Hochschule Hannover, Konstanty-Gutschow-Str.8, D-3000 Hannover 61, FRG

J.T.Gallagher
Department of Medical Oncology, University of Manchester, Christie Hospital and Holt Radium Institute, Wilmslow Road, Manchester M20 0BX, UK

M.P.Henry
J.T.Baker Inc., 222 Red School Lane, Phillipsburg, NJ 08865, USA

D.Johns
Technicol Ltd, Brook Street, Higher Hillgate, Stockport, Cheshire SK1 3HS, UK

R.W.A.Oliver
Biological Materials Analysis Research Unit, Department of Biological Sciences, University of Salford, Salford M5 4WT, UK

A.Pingoud
Abteilung Biophysikalische Chemie, Zentrum Biochemie, Medizinische Hochschule Hannover, Konstanty-Gutschow-Str.8, D-3000 Hannover 61, FRG

V.Pingoud
Abteilung Klinische Biochemie, Zentrum Laboratoriumsmedizin, Medizinische Hochschule Hannover, Konstanty-Gutschow-Str.8, D-3000 Hannover 61, FRG

G.W.Welling
Lab. voor. Med. Microbiologie, Rijksuniversiteit Groningen, Ostersingel 59, 9713 EZ Groningen, Holland

S.Welling-Wester
Lab. voor Med. Microbiologie, Rijksuniversiteit Groningen, Ostersingel 59, 9713 EZ Groningen, Holland

Contents

Abbreviations

AC	affinity chromatography
ADH	alcohol dehydrogenase
p-APM	*p*-aminophenyl-α-D-mannopyranoside
AUFS	absorbance units full scale
BSA	bovine serum albumin
CAGG	*N*-chloroacetylglycylglycine
CM	carboxymethyl
CDI	carbonyldiimidazole
Con A	concanavalin A
DEAE	diethylaminoethyl
DMF	dimethylformamide
DMSO	dimethylsulphoxide
ELISA	enzyme-linked immunosorbent assay
E-PHA	phytohaemagglutinin-erythrocytes
GalNAc	*N*-acetylgalactosamine
GlcNAc	*N*-acetylglucosamine
HA	hydroxyapatite chromatography
HFBA	heptafluorobutyric acid
HIC	hydrophobic interaction chromatography
HK	hexokinase
HPAC	high-performance affinity chromatography
HPIEC	high-performance ion-exchange chromatography
HPLC	high-performance liquid chromatography
HSA	human serum albumin
IEC	ion-exchange chromatography
L-PHA	phytohaemagglutinin-lymphocytes
LCA	lentil lectin (*Lens culinaris*) agglutinin
LDH	L-lactate dehydrogenase
α-mm	α-methyl mannoside
NeuAc	sialic acid (*N*-acetyl-neuraminic acid)
OPD	orthophenylene diamine dihydrochloride
PAGE	polyacrylamide gel electrophoresis
PBS	phosphate-buffered saline
PDGF	platelet-derived growth factor
PEI	polyethyleneimine
PGK	3-phosphoglycerate kinase
PNA	peanut agglutinin
QEA	quaternary amine
RCA	*Ricinius communis* agglutinin
RP-HPLC	reversed-phase HPLC
RPC	reversed-phase chromatography
SAX	strong anion-exchange
SDS	sodium dodecyl sulphate
SE	size-exclusion
SEC	size-exclusion chromatography
SE-HPLC	size-exclusion HPLC
TAPA	trimethyl (*p*-aminophenyl) ammonium chloride

TCA	trichloroacetic acid
TEAP	triethylammonium phosphate
TFA	trifluoroacetic acid
TSH	thyrotropin
WGA	wheat germ agglutinin

Pressure units

1 bar $\equiv$ 14.5 p.s.i. $\equiv$ 0.1 MPa.

CHAPTER 1

Columns for HPLC separation of macromolecules

DENISE JOHNS

1. INTRODUCTION

The application of high-performance liquid chromatography (HPLC) to the separation of macromolecules is a fairly recent development. Traditionally, protein chemists have used semi-rigid or non-rigid gels which have a relatively low mechanical strength, limiting the chromatographer to low flow-rates and large particle sizes, and resulting in poor resolution with long separation times. Alternative techniques, such as electrophoresis consequently prevailed.

HPLC was limited by the slow development of suitable alternative stationary phases, but as a result of the last few years progress, the chromatographer is now presented with a bewildering choice of both techniques and stationary phases. This chapter will try to unfold the seemingly complex nature of modern HPLC support materials by explaining the reasons for their development and the differences between the major phases available today. Because different types of biological macromolecules utilize varying types of chromatographic technique, proteins and peptides, oligonucleotides, immunoglobulins and oligosaccharides will be discussed separately.

2. COLUMNS AND PACKING MATERIALS FOR PROTEIN AND PEPTIDE HPLC

What is required for the separation of proteins and peptides is a column support material that will withstand high pressures, allowing eluants to be pumped through the column at high velocities, and one which avoids the polar, highly absorptive surface associated with many untreated silica based HPLC phases.

In summary, a support for the separation of these biopolymers ideally needs to be:

(i)   chemically stable, but not unreactive such that chemical modification is difficult;
(ii)   macroporous, to allow high-molecular-weight proteins to permeate and make contact with the media surface, and yet rigid enough to withstand the operating pressures encountered in HPLC;
(iii)   well endowed in surface area, for best possible resolution characteristics;
(iv)   weakly hydrophilic, to allow wetting of the surface, but not soluble in water;
(v)   if the surface of the support requires modifying, or bonding, in order to achieve these criteria, such bonding needs to be highly selective so as to minimize non-specific adsorption.

Several approaches have been made to solve these seemingly insurmountable criteria and many are highly successful. The fact that proteins vary enormously in their behaviour

1

can be utilized for their separation in HPLC. The diverse biological and biochemical functions of proteins are largely due to the different amino acid side chains which convey different properties onto the protein, providing a guide to possible methods of separation.

(i)   Interaction between the side chains of a protein or peptide results in a specific (tertiary) structure for the protein and hence differences in size can be utilized for HPLC [size-exclusion chromatography (SEC), or SE-HPLC].

(ii)   Amino acid side chains may be ionized, either as anions or cations and this ionization will change with pH (making ion-exchange chromatography a useful mode to use).

(iii) Amino acid side chains may also differ in their polarity: reversed-phase chromatography is therefore a popular technique.

(iv)   Because of their different conformational states, proteins will bind specifically to certain molecules, for example enzymes will bind to substrates, co-enzymes, inhibitors and activators. If these are immobilized on an HPLC column, separation may be achieved (affinity chromatography).

The availability of column supports which utilize each of these four HPLC modes for the separation of proteins and peptides is discussed in Sections 2.1–2.4.

2.1 Size-exclusion HPLC

Although high-performance columns for size-exclusion have been available for some time, the significant development for protein chemists has been the recent introduction of rigid aqueous compatible SEC materials. These allow the separation of biopolymers by molecular size, the largest molecules eluting from the column first.

2.1.1 *Choice of SEC column*

Stationary phases for SEC achieve separation by simply depending on the size of the molecules affecting the degree of penetration into the pores. The pore size is therefore chosen such that small molecules pass in and out of the pores essentially without hindrance, whilst large molecules which do not penetrate the pores, elute from the column first at the void volume. Molecules of intermediate size, about the size of the pore, pass through the column with difficulty, resulting in differing retention times. Some columns, notably silica gel columns (1,2), have a high enough resolution to separate molecules which have small size differences, even for molecular weights below 1000.

The initial task in using these columns is to choose the appropriate pore size for the separation. In practice, all manufacturers quote a molecular weight exclusion limit, above which a protein molecule will be totally excluded, and they may also provide a calibration curve which plots (i) the logarithm of molecular weight against elution volume (*Figure 1*) or (ii) the logarithm of molecular weight against the distribution coefficient K_D (*Figure 2*). K_D is defined as the ratio

$$\frac{V_e - V_o}{V_t - V_o}$$

where V_e = elution volume of a macromolecule, V_o = void volume of the column,

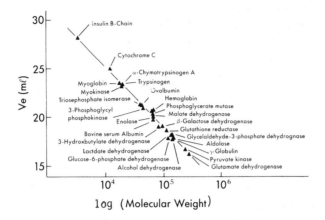

Figure 1. Calibration curve for proteins on TSK 3000SW columns. Column, TSKgel G3000SW (7.5 mm i.d. × 60 cm); eluant, 50 mM phosphate buffer, pH 7.5 + 0.2 M NaCl; flow-rate, 0.3 ml/min.

V_t = elution volume of small molecule. The standards used to derive these calibration curves are frequently dextrans or sulphonated polystyrenes, although this tends to (incorrectly) assume a linear structure for the biological macromolecule. The behaviour of proteins and peptides on SEC columns is more accurately represented by the use of globular proteins as calibration standards. Anderson and Stoddart observed (3) that in the middle of the K_D range ($K_D = 0.15 - 0.80$) these theoretical plots are essentially linear (see *Figure 2*), and they can therefore assist in the choice of SEC column. It is consequently generally best (4) to choose a column on which the molecules of interest elute near the middle of the calibration curve, where this linearity occurs. Occasionally, two or more columns may be used in series. When connected in series, columns of the same pore size will increase resolution, whilst columns of different pore size will broaden the range of molecular sizes separated.

The choice of SEC column must also take into account whether the protein is to be denatured before analysis. It is common to employ denaturing mobile phases, such as those containing detergents or organics, particularly when using silica gels (which are very stable in the presence of denaturants). The use of sodium dodecyl sulphate, urea, guanidinium chloride, or organic solvents such as acetonitrile, will denature the protein, preventing self-association, and increasing solubility. This can, however, greatly decrease the radius of the protein, and their exclusion range will move to lower molecular weights. *Table 1* shows reported exclusion limits for SEC columns of different pore sizes; denatured proteins will exhibit calibration curves more closely resembling linear molecules than those of globular proteins.

One final criteria for the choice of SEC column is that there should be no interaction between the solute and the surface of the stationary phase so that there is nothing eluting after the smallest molecule, that is separation occurs exclusively by size-exclusion. The problem of excessive interaction of the protein with the stationary phase, especially common with silica based materials, is solved by bonding the surface with a modifier. However, bonding on silica is often incomplete for steric reasons, so these materials

3

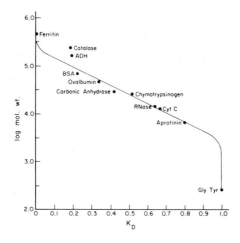

Figure 2. Calibration curve for proteins on SynChropak 100 (5 μm).

Table 1. Reported molecular weight exclusion limits for linear and globular macromolecules on differing pore size SEC supports manufactured by SynChrom Inc. (SynChropak).

Pore size of SEC support (nm)	Molecular weight			
	Linear molecule		Globular molecule	
	Excluded	Included	Excluded	Included
6	20 000	300	30 000	300
10	60 000	2000	300 000	3000
30	200 000	4000	1 500 000	6000
50	300 000	9000	10 000 000	20 000
100	400 000	15 000	50 000 000	25 000
400	10 000 000	20 000	Not available	

often exhibit residual ion-exchange properties. The effects of the slightly negative charge on residual silanols can be compensated for by adding salts to the aqueous mobile phase. Generally, a 0.05−0.2 M solution of buffer (e.g. potassium phosphate) is effective, although care should be taken not to use too high a salt concentration since this increases hydrophobic interactions. If these occur, an organic modifier, such as methanol, can be added, or an alternative mobile phase could be used (e.g. 0.1% v/v trifluoroacetic acid for the elution of peptides).

2.1.2 Commercially available SEC materials

Columns containing support materials having pore sizes from 5 to 400 nm are readily available. *Table 2* lists some of those available commercially and includes both silica based and crosslinked polystyrene (polymeric) supports. The manufacture of SEC materials having a carefully controlled pore size and pore size distribution is accomplished in several ways. For polymeric materials, there are basically two methods of manufacture. By carefully controlling the crosslinking in a polystyrene−divinyl

Table 2. Alphabetical name listing of column support materials available for SE-HPLC of proteins and peptides.

Support	Pore size (nm)	Mol. wt exclusion limit for globular proteins	Particle size (µm)	Column sizes Length (cm)	i.d. (mm)	Chemical nature	Availability bulk material	Manufacturer[a]
Aquapore OH-100	10	9×10^4	10	25	4.6	Silica	No	Brownlee
Aquapore OH-300	30	8×10^5	10	25	4.6	Silica	No	Brownlee
Aquapore OH-500	50	5×10^6	10	25	4.6	Silica	No	Brownlee
Aquapore OH-1000	100	20×10^6	10	25	4.6	Silica	No	Brownlee
LiChrospher DIOL	10	7×10^4	5,7	25	4.0	Silica	Yes	E.Merck
LiChrospher DIOL	50	8×10^5	10	25	4.0	Silica	No	E.Merck
LiChrospher DIOL	100	3×10^6	10	25	4.0	Silica	No	E.Merck
MCI Gel CQS10	10	4×10^5	9–11	30	7.5	Polymer	Yes	Mitsubishi
MCI Gel CQS30	30	1×10^6	9–11	30	7.5	Polymer	Yes	Mitsubishi
MCI Gel CQP06	6	1×10^{3b}	9–11	30	7.5	Polymer	Yes	Mitsubishi
MCI Gel CQP10	10	4×10^5	9–11	30	7.5	Polymer	Yes	Mitsubishi
MCI Gel CQP30	20	1×10^6	7–10	30	7.5	Polymer	Yes	Mitsubishi
Polyol Si300	30	6×10^5	3,5,10	25,50	4.6,7.1,9.5	Silica	Yes	Serva
Polyol Si500	50	1×10^6	10	25,50	4.6,7.1,9.5	Silica	Yes	Serva
SynChropak 60	6	3×10^4	5	25,30	4.6,7.8	Silica	Yes	SynChrom
SynChropak 100	10	3×10^5	5	25,30	4.6,7.8	Silica	Yes	SynChrom
SynChropak 300	30	1.5×10^6	5	25,30	4.6,7.8	Silica	Yes	SynChrom
SynChropak 500	50	10×10^6	7	25,30	4.6,7.8	Silica	Yes	SynChrom
SynChropak 1000	100	50×10^6	7	25,30	4.6,7.8	Silica	Yes	SynChrom
SynChropak 4000	400	NA	10	25,30	4.6,7.8	Silica	Yes	SynChrom
SOTA GF200	20	7×10^5	10	30	7.1	Silica	No	Sota
TSK 2000 SW	12.5	7×10^4	10	30,50	7.5,21.5	Silica	No	Toya Soda
TSK 3000 SW	25	3×10^5	10	30,60	7.5,21.5	Silica	No	Toya Soda
TSK 4000 SW	50	1×10^6	13	30	7.5	Silica	No	Toya Soda
TSK 2000 PW	5	4×10^{3b}	10	30,50	7.5,21.5	Polymer	No	Toya Soda
TSK 3000 PW	20	5×10^{4b}	13	30,60	7.5,21.5	Polymer	No	Toya Soda
TSK 4000 PW	50	3×10^{5b}	13	30,60	7.5,21.5	Polymer	No	Toya Soda
TSK 5000 PW	100	1×10^{6b}	17	30,50	7.5,21.5	Polymer	No	Toya Soda
TSK 6000 PW	>100	8×10^{6b}	25	30,50	7.5,21.5	Polymer	No	Toya Soda
Zorbax Bio GF-250	25	4×10^5	4	25	9.4	Silica	No	Du Pont

NA = not available.
[a]See Appendix for complete names and addresses of manufacturers.
[b]PEG standard.

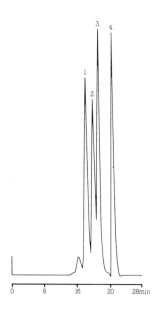

Figure 3. The size-exclusion chromatographic separation of a standard protein mixture on a Hibar RT LiChrospher 500 DIOL column. Column size, 250 × 4 mm (2 columns); eluant, 50 mM H_3PO_4 adjusted to pH 7.0 with 2.0 M NaOH + 1% SDS; flow-rate, 0.2 ml/min; pressure, 15 kg/cm^2; chart speed, 2.5 mm/min; detection, UV 208 nm 0.08 AUFS. Samples, (**1**) albumin (bovine serum = 68 000); (**2**) chymotrypsinogen A (25 000); (**3**) cytochrome *c* (12 500); (**4**) alanine (89).

benzene co-polymer, a highly rigid material with a specific pore size can be produced. These microporous SEC materials, as they are known, generally tend to have irregular particles, and the control of crosslinking in this way is critical to achieve an optimum efficiency plus optimum rigidity. The production of macroporous materials, on the other hand, makes use of a poragen, a substance which is insoluble in the polymer but soluble in the monomer such that it washes out at the end of the manufacturing process leaving regular-sized pores. Silica based SEC materials are manufactured by carefully controlling the initial stages of precipitation of the silicate to produce a macroporous material of the required pore volume. Manufacture of the larger pore sized silica materials (100 nm and above) at first proved difficult, the porous particles suffering collapse in use: recent production methods seem to have generally overcome these problems. There is also a method for preparing glass beads with uniform pore sizes for SEC work (5), but these have been largely superceded by polymeric supports for protein work.

Several different approaches to the bonding of silicas for SEC work have been developed. Diol modified silicas (6–8), produced via 1,2-dihydroxy-3-propoxypropyl bonding (Merck, Darmstadt), have been successfully used in the separation of proteins (*Figure 3*), although the elution behaviour has been found to be influenced to a small extent by ionic interaction due to unreacted silanols (6). Stabilizing the surface of a diol bonded 4 μm spherical silica with metal oxides has produced an SEC material which can operate at pH values up to 8.5, for example Zorbax GF-250 (Du Pont UK Ltd, Stevenage, UK) utilizes zirconia treatment to provide high efficiencies and high sample recoveries (typically greater than 95%).

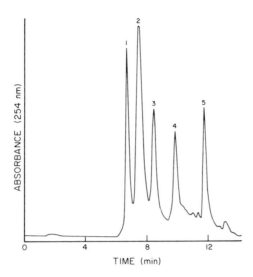

Figure 4. SE-HPLC of a mixture of proteins on SynChropak 100. Column, 250 × 4.6 mm i.d.; mobile phase, 0.1 M potassium phosphate, pH 7; flow-rate, 0.25 ml/min. Sample, (**1**) thyroglobulin; (**2**) alcohol dehydrogenase; (**3**) ovalbumin; (**4**) ribonuclease; (**5**) glycyl tyrosine.

SynChropak SEC supports (SynChrom Inc., Lafayette, USA), and also Aquapore OH columns (Brownlee Labs, USA), are composed of silica with a glycerylpropyl bonded surface and are available in six pore diameters, enabling the separation of a wide range of proteins and peptides (*Figure 4*). Their capacity is good for proteins: generally 200−400 µg of protein can be loaded onto a 25 cm × 4.6 mm column before band spreading occurs, while a 30 cm × 7.8 mm semi-preparative column can handle 2−4 mg.

The Toyo Soda Company (Tokyo, Japan) manufactures a range of both silica based and polymeric columns which are probably the most commonly known SEC materials. The TSKgel SW type is based on a hydrophilic rigid spherical silica suitable for the analysis of proteins and enzymes because of its low adsorption and narrow pore size distribution (*Figure 1* shows a typical calibration curve for one of the SEC columns). The corresponding polymeric columns comprise the TSKgel PW series but these are more commonly used for oligosaccharides and oligonucleotides.

A new material based on a particle size 10 µm, pore size 20 nm silica coated with a hydrophilic bonded phase is SOTA Phase GF200 (SOTA Chromatography Inc., Crompond, USA). When used with 100−150 mmol salts at neutral or slightly acidic pH, GF200 columns, like the Bio-Sil TSK series, provide high recoveries of proteins, and in most cases biological activity is retained. *Figure 5* compares similar separations on both columns.

2.2 Ion-exchange chromatography

2.2.1 *Choice of IEC column*

Because a protein can exist either as an anion or as a cation, ion-exchange

7

chromatography provides a custom-designed system for the separation of proteins. Retention will depend on the competition between the protein ionic sites and the mobile phase for the ionic sites on the support. The type of ionic sites on the protein will vary with pH, and their number will vary with the protein type and its tertiary structure (polar moieties will often be effectively shielded). For peptides with marked differences in ionic character, ion-exchange chromatography will probably provide the greatest selectivity. Angiotensin is an example of a peptide which chromatographs well on ion-exchange. If a range of peptides to be analysed are similar ionically, but differ in their hydrophobicity, it is likely that reversed-phase HPLC will be a better choice (see Section 2.3).

Competition of the protein with the mobile phase will vary with the eluant pH and ionic strength, and therefore elution of proteins can be brought about by a gradient of increasing ionic strength, a pH gradient, or a combination of both (less common). Ion-exchange chromatography consequently provides the protein and peptide chemist with a wide range of variables for influencing retention and selectivity: this is one of the reasons for its popularity as an HPLC technique, although it can often be the hardest method to develop.

2.2.2 *Commercially available IEC materials*

Both silica and polymeric based columns are regularly used for ion-exchange

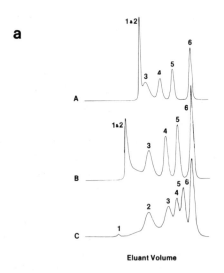

Figure 5 (a). SE-HPLC separation of large mol. wt proteins on Bio-Sil TSK. Instrument, Bio-Rad's protein chromatography system; column, **(A)** Bio-Sil TSK 125, 300 × 7.5 mm; **(B)** Bio-Sil TSK 250, 300 × 7.5 mm; **(C)** Bio-Sil TSK 400, 300 × 7.5 mm; sample, Bio-Rad's gel filtration standard, 20 μl; eluant, 0.1 M Na$_2$SO$_4$, 0.02 M NaH$_2$PO$_4$, pH 6.8; flow-rate, 1.0 ml/min; detection, UV 280 nm. Peaks, **(1)** protein aggregates (void peak); **(2)** thyroglobulin (670 000); **(3)** gamma globulin (158 000); **(4)** ovalbumin (44 000); **(5)** myoglobin (17 000); **(6)** vitamin B-12 (1350).

b

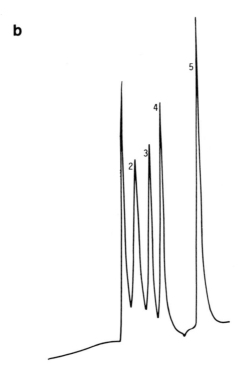

Figure 5 (b). SE-HPLC separation of large mol. wt proteins on SOTA GF200 columns. Column size, 300 × 7.1 mm; eluant 0.1 M Na_2SO_4, 0.02 M NaH_2PO_4, pH 6.9; flow-rate, 0.4 ml/min; detection, UV 280 nm. Peaks, (**1**) thyroglobulin; (**2**) gamma globulin; (**3**) ovalbumin; (**4**) myoglobin; (**5**) vitamin B-12.

chromatography and *Table 3* gives an indication of the wide choice commercially available. Silica gel supports have been more frequently used, being available with a wide variety of subtly different bonding chemistries. Recently, however, hydrophilic organic polymers, such as the hydroxylated polyether based support (TSK PW), have become available and provide good results for both analytical and preparative separations. One advantage that polymeric materials have over silica media for this work is that their stability at high pH enables the use of more stringent column cleaning procedures (e.g. the use of strong bases to remove over-retained components), increasing the working life of the column.

Silica gels are often bonded with strongly acidic or strongly basic moieties which enable desorption from the column in a much smaller elution volume, thereby increasing sensitivity and resolution. Further, with strong ion-exchangers, the pH of the mobile phase does not affect the ionization of the surface, making optimization a little easier (9). Many are available as wide pore gels, an advantage for high-molecular-weight proteins because with greater accessibility to ion-exchange sites, the capacity of columns is increased. A pore size of 30 nm is recommended for proteins except for macro-molecules with molecular weights over 150 000, when a pore size of 100 nm is better.

Weak anionic and cationic ion-exchange materials are also frequently used for peptides

Table 3. Some commercially available column support materials for ion-exchange HPLC of proteins.

	Surface	Functional group	Pore size (nm)	Particle size (µm)	Column dimensions length (cm) × i.d. (mm)	Manufacturers
Silica based columns						
Aquapore AX-300	Weak anion	DEAE	30	10	10,22 × 4.6, 25 × 7.0	Brownlee
Aquapore AX-1000	Weak anion	DEAE	10	10	10,22 × 4.6, 25 × 7.0	Brownlee
Aquapore CX-300	Weak cation	Carboxymethyl	30	10	10,22 × 4.6, 25 × 7.0	Brownlee
Aquapore CX-1000	Weak cation	Carboxymethyl	10	10	10,22 × 4.6, 25 × 7.0	Brownlee
Bakerbond PEI	Weak anion	Polyethyleneimine, CH_2CH_2NH	30	5	5 × 4.6	J.T.Baker
Bakerbond CBX	Weak cation	Carboxyethyl	30	5	5 × 4.6	J.T.Baker
SynChropak AX300	Weak cation	Polyamine	30	6.5	10,25 × 4.6,10,22.5	SynChrom Inc.
SynChropak AX1000	Weak anion	Poly	30	10	10,25 × 4.6,10,22.5	SynChrom Inc.
SynChropak Q300	Strong anion	Quaternary amine	30	6.5	10,25 × 4.6,10,22.5	SynChrom Inc.
SynChropak Q1000	Strong anion	Quaternary amine	30	10	10,25 × 4.6,10, 22.5	SynChrom Inc.
SynChropak S300	Strong cation	Sulphonic acid	30	6.5	10,25 × 4.6,10	SynChrom Inc.
SynChropak CM300	Weak anion	Carboxymethyl	30	6.5	10,25 × 4.6,10	SynChrom Inc.
TSK DEAE-2/SW	Weak anion	$-N^+HEt_2$	13	5	30 × 4.6	Toya Soda
TSK DEAE-3/SW	Weak anion	$-N^+HEt_2$	25	10	15 × 6.0,21.5	Toya Soda
TSK CM-2/SW	Weak cation	-COO	13	5	30 × 4.6	Toya Soda
TSK CM-3/SW	Weak cation	-COO-	25	10	15 × 6.0,21.5	Toya Soda
TSK CM-5/SW	Weak cation	-COO-	50	10	15 × 4.6,21.5	Toya Soda
Waters Accell	Cation	Carboxymethyl	50	37–55	Bulk material only	Waters
Polymer based columns						
Mono Q	Anion	Quaternary amine	NA	10	50 × 5	Pharmacia
Mono S	Cation	Sulphonate	NA	10	50 × 5	Pharmacia
PL SAX	Strong anion	Quaternary amine	100	8,10	5,15 × 4.6,5,15 × 7.5,25	Polymer
TSK DEAE 5 PW	Weak anion	$-N^+HEt_2$	100	10	7.5 × 7.5,15 × 21.5	Toya Soda
TSK SP 5 PW	Strong cation	$-SO_3^-$	100	10	7.5 × 7.5,15 × 21.5	Toya Soda
TSK Gel DEAE-NPR	Weak anion	$-N^+HEt_2$	100	2.5	3.5 × 4.6	Toya Soda
TSK Gel SP-NPR	Weak cation	$-(CH_2)_3SO_3^-$	100	2.5	3.5 × 4.6	Toya Soda

NA = not available.

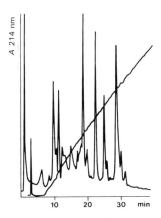

Figure 6. Fractionation of tryptic peptides of carboxymethylated 1,4-glucan cellobiohydrolase from *Trichoderma reesei* on Mono Q, monitored at 214 nm. Sample, 80 min tryptic digest; flow-rate, 1 ml/min; eluant, 20 mM Tris−HCl, pH 8.0 with NaCl gradient 2.5 mM−0.5 M. (From Pharmacia Fine Chemicals AB, Uppsala, Sweden.)

and proteins. Bonding silica with diethylaminoethyl (DEAE) or carboxymethyl (CM) moieties gives materials which are chemically stable without unwanted hydrophobic interactions. The TSK SW range is available in 13 nm pore size for small proteins over 20 000 molecular weight. Typically, ion-exchange capacities are greater than 0.3 meq/g. For preparative HPLC, a 50 nm weak cation-exchanger (Accell CM) is available in 37−55 μm particle size (Millipore UK Ltd). It is based on a rigid non-compressible silica support.

A particularly efficient way of bonding silica gel with an ion-exchanger to minimize the number of non-specific sites is achieved by crosslinking the ionic ligands within a thin polymer film. The SynChropak AX300 weak anion-exchanger uses a polyamine polymer coating to ensure complete coverage. Elution protocols and recoveries are similar to those obtained with DEAE gels.

Elimination of the usual pH dependence of silica based ion-exchangers has been achieved in the SynChropak series of columns by alkylating the AX300 to produce a strong anion-exchanger (Q300 and Q1000) similar in selectivity to QEA (quaternary amine) celluloses. Additionally, a strong cation-exchanger (SynChropak S300) produced by forming a polymeric layer of sulphonic acids has been shown to be useful for basic proteins and peptides, with elution behaviour close to that expected for propylsulphonate gels.

Of the polymeric phases available for ion-exchange chromatography of proteins, the Mono Q and Mono S resins (Pharmacia Fine Chemicals) are probably the most prolific. They are macroporous, hydrophilic particles produced from a polystyrene−divinyl benzene co-polymer as a monosized 10 μm particle. The low dispersion enables high flow-rates with low back-pressures, while a very low metal ion concentration ensures good peak shape. They have high mechanical stability and do not suffer from the swelling often associated with polymeric materials. Fast separations with high recovery are frequently achieved, one example being the fractionation of the tryptic peptides of

11

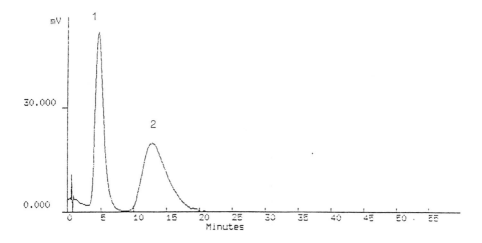

Hormone	R_t (mins)	% Recovery
Angiotensin I	12.1	100
Angiotensin II	4.4	100

Figure 7. Separation of two angiotensins on preparative column of Dyno particles (PD-201-SCX). Column, PD-201-SCX (4.6 mm × 25 cm); mobile phase, 0.2 M sodium phosphate pH 6.7/CH_3CN (70/30); flow-rate, 4.0 ml/min; detection, UV 220 nm; temperature, room temperature. (**1**) Angiotensin I, Asp-Arg-Val-Tyr-Ile-His-Pro-Phe-His-Leu; (**2**) angiotensin II, Asp-Arg-Val-Tyr-Ile-His-Pro-Phe.

cellulase from *Trichoderma reesei* (*Figure 6*). For the large scale preparative and process HPLC of peptides, a similar macroporous monosized polymeric strong cation-exchanger is also available (Dyno Particles, Norway). Angiotensins containing amino acids with aromatic side chains are often difficult to elute from ion-exchange supports, but this material has been reported to exhibit excellent recovery and separation for two angiotensins (*Figure 7*).

The introduction of diethylaminoethyl and propylsulphonic acid groups onto a non-porous hydrophilic polymer of 2.5 μm in particle diameter produces a pair of materials that can rapidly separate proteins with high resolution and high recovery. The TSKgel DEAE−NPR and TSKgel SP−NPR resins have an ion-exchange capacity of 0.15 meq/ml of column bed volume and provide an addition to the family of TSK resins, the TSKgel DEAE−5PW, and SP−5PW (*Table 3*), now well known for their application in protein and enzyme separations.

Another macroporous polymeric material recently introduced is a 100 nm strong cation-exchanger (Polymer Laboratories). They report that by utilizing a large pore size of 100 nm, the maximum mass transfer, loading and resolution for proteins is possible (*Figure 8*), while the polymeric matrix ensures minimal non-specific sites.

2.3 Reversed-phase chromatography

The reversed-phase (RP)-HPLC of proteins and peptides has in recent years proved to be one of the most efficient and versatile methods of separation, especially for closely

12

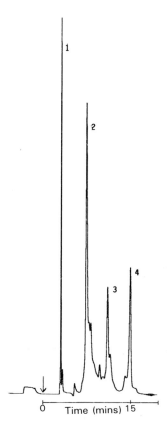

Figure 8. Ion-exchange HPLC of a mixture of proteins. Column, PL-SAX 100 nm 8 μm 50 $\times$ 4.6 mm; eluant A, 0.01 M Tris–HCl, pH 8.0; eluant B, 0.01 M Tris–HCl, 0.35 M NaCl, pH 8.0; gradient, linear 0–100% B over 20 min; flow-rate, 1.0 ml/min; detector, UV at 280 nm. Peaks, (**1**) myoglobin; (**2**) conalbumin; (**3**) ovalbumin; (**4**) soyabean trypsin inhibitor.

related molecules. Examples of the extensive list of applications include vasopressin and oxytocin (peptide hormones), cytochrome *c*, endorphins and tryptic digests of proteins. It is a form of hydrophobic interaction chromatography in which the stationary phase is rather more non-polar than is usual. It also differs from conventional reversed-phase HPLC in that the supports generally have much larger pores than usual to accommodate the larger molecular sizes encountered in the separation of biological macromolecules. Typically, a silica gel that has been chemically bonded with octadecylsilane for RP-HPLC can have as much as 80% of the total ODS inside the pore. To maximize the possibility of separation, it is therefore important for the macromolecule to have access to as much of the surface as possible. The use of 30 nm reversed-phase silicas for protein and peptide HPLC has therefore become commonplace, this pore size offering an optimum in rigidity and applicability to a wide range of proteins.

2.3.1 *Choice of RP column*

The hydrophobic amino acids in a peptide or protein retain on a reversed-phase surface

13

by single, or multi-point, adsorption. It has been reported (10,11) that, because of this interaction, the retention time of a protein or peptide can be predicted fairly accurately from the ratio of polar to non-polar amino acids in the molecule. This will, of course, be complicated by the tertiary structure of the molecule, which may even be different in the mobile phase, making some amino acids unavailable for interaction.

The separation of proteins in reversed-phase HPLC occurs because of the competition between the non-polar stationary phase, the non-polar sites of the protein, and the organic modifier of the mobile phase (which generally has the lowest non-polar character of the three). In contrast to reversed-phase HPLC of small molecules, which involves adsorption and partition, proteins and peptides interact with the surface of the support by an adsorption type mechanism which is highly dependent on conformation.

Geng and Regnier (12) have proposed a mechanism for the reversed-phase absorption of proteins whereby a protein, upon entering the column, displaces the mobile phase organic modifier on the surface of the reversed-phase support. The protein then remains adsorbed to the support until a certain 'critical' concentration of organic modifier is achieved. The protein is then desorbed from the support by a certain number of organic modifier molecules. Geng and Regnier designate the number of organic modifier molecules required to desorb the protein as the 'Z number'. This 'Z number' is a characteristic of the protein or peptide being chromatographed. It is dependent on the organic modifier used as well as the degree of denaturation of the protein.

Since the protein is desorbed quickly upon reaching the 'critical concentration' of organic modifier, in other words when the proper 'Z number' is reached, the capacity factor tends to change very rapidly with changes in organic modifier percentage. If the solute capacity factor k' [defined as $k' = (t_R t_S)/t_S$ where t_R = retention time, t_S = the value of t_R for the solute when it is unretained by the column] is plotted versus the percentage of organic modifier (for example, acetonitrile), the plots drop very sharply when the 'critical concentration' is reached and then remain nearly constant at a very low value as higher organic modifier concentrations are reached. This may be compared with the much more gentle slope of the plot of the capacity factor versus acetonitrile concentration found with a small molecule such as biphenyl.

Advantage can be made of the fact that salt-free eluants are used in reversed-phase HPLC to facilitate their use in preparative HPLC. A solvent that is frequently used is 0.1% trifluoroacetic acid (TFA) which is freely available at the high purity required for UV monitoring at 215 nm, and which readily evaporates. Additionally, being a strong acid (0.1% TFA gives a pH of about 2), it is an excellent solubilizing agent. Even at low concentrations, it will protonate the carboxyl groups of peptides, increasing the affinity of the peptide for the reversed-phase surface. Organic modifiers are usually acetonitrile or isopropanol, acetonitrile being more hydrophobic than water but less hydrophobic than isopropanol, which is therefore more useful for eluting larger, more hydrophobic proteins (usually replacing the acetonitrile with a 2:1 mixture of acetonitrile to isopropanol).

Mass recovery of proteins and peptides is generally good on reversed-phase HPLC. Most peptides are recovered at 90−100% while even very basic proteins such as histones show recoveries of at least 80%.

It is well known that the conditions of reversed-phase chromatography normally disrupt the tertiary structure of proteins and cause loss of biological activity. Further, peptides and proteins which do not appear to be hydrophobic may behave as such on reversed-phase because of conformational changes. Proteins may also aggregate during reversed-phase chromatography, causing over-long retention times, although this can be rectified by the use of 3 M guanidine or 3 M urea in 50% aqueous isopropanol. In many cases, however, some or all of the biological activity may be recovered by attention to certain factors.

(i) Use of alcohol modifiers, especially isopropanol, usually results in higher biological activity recovery.

(ii) Short residence times on the column favour recovery of activity since denaturation is a kinetically slow process.

(iii) The presence of co-factors, especially if added after collection of the protein fraction, enhances the recovery of activity.

(iv) The treatment of the collected fraction after chromatography may be the most important factor affecting the recovery of activity. Gentle treatment, addition of buffers or co-factors, incubation time and low temperatures may do much to allow recovery of activity.

(v) Initial solvent conditions, especially the organic modifier concentration, may affect the biological activity.

Even with these apparent difficulties, reversed-phase HPLC remains one of the most important techniques for the chromatography of proteins.

2.3.2 *Commercially available RP columns*

Only those silica based reversed-phases with as few as possible residual silanols are suitable for the HPLC of peptides and proteins. The effect of the silanols usually remaining after bonding and often remaining even after end-capping with trimethylsilane (TMS) is to increase the retention time of a protein and peptide molecule. This may also lead to asymmetric peaks, and in some cases to poor recovery and inaccurate quantitative results. To a certain extent, the problem can be overcome by the addition of ionic components to the mobile phase (another reason for using TFA), or by keeping the pH to below 4, when the silanols are largely associated.

Conversely, this problem has led to a new generation of reversed-phases specially developed for protein and peptide HPLC and these are summarized in *Table 4*. The use of wide pore silica bases is common, being better suited to the larger molecular weights of these compounds, as explained earlier. Bonding these with octadecylsilane and end-capping with TMS will often protect the silanols well enough, partly because of steric hindrance, and partly because of the lower surface area of such materials. For the separation of proteins and peptides, however, further steps need to be taken to ensure full recovery of the compounds of interest. Special manufacturing processes are employed by one producer (The Separations Group) to produce silicas which have a low metal and sulphate content. The presence of significant amounts of these ions, often found in general purpose column supports for reversed-phase chromatography,

Table 4. Summary of the commercially available column support materials for reversed-phase HPLC of proteins and peptides.

Support	Surface	Pore size (nm)	Particle size (μm)	Column dimensions length (cm) × i.d. (mm)	Bulk material available	Manufacturers
Apex WP Octadecyl	C_{18}	30	7	25,10 × 4.6	Yes	Jones Chromatography Ltd
Apex WP Octyl	C_8	30	7	25,10 × 4.6	Yes	Jones Chromatography Ltd
Apex WP Butyl	C_4	30	7	25,10 × 4.6	Yes	Jones Chromatography Ltd
Apex WP Phenyl	Phenyl	30	7	25,10 × 4.6	Yes	Jones Chromatography Ltd
Aquapore RP-300	C_8	30	7	22,10,30 × 4.6,2.1	No	Brownlee
Aquapore RP-300	C_4	30	7	22,10,30 × 4.6,2.1	No	Brownlee
Aquapore PH-300	Phenyl	30	7	22,10,30 × 4.6,2.1	No	Brownlee
Bakerbond WP Octadecyl	C_{18}	30	5	25,10 × 4.6	Only 40 μm	J.T.Baker
Bakerbond WP Octyl	C_8	30	5	25 × 4.6	Only 40 μm	J.T.Baker
Bakerbond WP Butyl	C_4	30	5	25 × 4.6	Only 40 μm	J.T.Baker
Bakerbond WP Diphenyl	Diphenyl	30	5	25 × 4.6	Only 40 μm	J.T.Baker
Hypersil WP 300-Octyl	C_8	30	5,10	25,10 × 4.6,7	Yes	Shandon Southern
Hypersil WP 300-Butyl	C_4	30	5,10	25,10 × 4.6,7	Yes	Shandon Southern
PLRP-S-300[a]	C_{18}	30	8,10	25,10,50 × 4.6,10,25	No	Polymer Laboratories
PLRP-S-1000[a]	C_{18}	100	8,10	15,50 × 4.6,7.5	No	Polymer Laboratories
Serva Octadecyl=Si100	C_{18}	10	5,10	25 × 4.6	Yes	Serva GmbH

Serva Octadecyl=Si300	C$_{18}$	30	5,10	25 × 4.6	Yes	Serva GmbH
Serva Octadecyl=Si500	C$_{18}$	50	10	25 × 4.6	Yes	Serva GmbH
Serva Octyl=Si100	C$_8$	10	5,10	25 × 4.6	Yes	Serva GmbH
SynChropak RP-18	C$_{18}$	10,20,100	6.5	5,10,25 × 4.6,10,22.4	Yes	SynChrom Inc.
SynChropak RP-8	C$_8$	30,100	6.5	5,10,25 × 4.6,10,22.4	Yes	SynChrom Inc.
SynChropak RP-4	C$_4$	30,100	6.5	5,10,25 × 4.6,10,22.4	Yes	SynChrom Inc.
SynChropak RP-1	C$_1$	100	6.5	5,10,25 × 4.6,10,22.4	Yes	SynChrom Inc.
TSKgel ODS-80TM	C$_{18}$	8	5	25 × 4.6	Yes	Toya Soda Co.
TSKgel ODS-120A	C$_{18}$	12	5,10	25 × 4.6,7.8,21.5	No	Toya Soda Co.
TSKgel ODS-120T	C$_{18}$	12	5,10	25 × 4.6,7.8,21.5	No	Toya Soda Co.
TSKgel TMS-250	C$_1$	25	10	7.5 × 4.6	No	Toya Soda Co.
TSKgel Phenyl-5PN	Phenyl	NA	10	7.5 × 4.6	No	Toya Soda Co.
Ultrapore RPSC	C$_3$	30	5	7.5 × 4.6	No	Beckman Inc.
Ultrasphere ODS	C$_{18}$	5–9	5	25,15 × 4.6,10	No	Beckman Inc.
Ultrasphere ODS	C$_{18}$	8–9	3	7.5 × 4.6	No	Beckman Inc.
Vydac 218TP	C$_{18}$	30	5,10	25,15 × 4.6,2.1	Yes (10 μm only)	The Separations Group
Vydac 214TP	C$_4$	30	5,10	25,15 × 4.6,2.1	Yes (10 μm only)	The Separations Group
Vydac 219TP	Diphenyl	30	5,10	25,15 × 4.6,2.1	Yes (10 μm only)	The Separations Group
Vydac 228TP	C$_8$	3	5,10	25,15 × 4.6,2.1	Yes (10 μm only)	The Separations Group

[a]Polymeric column support; all others are silica based.
NA = not available.

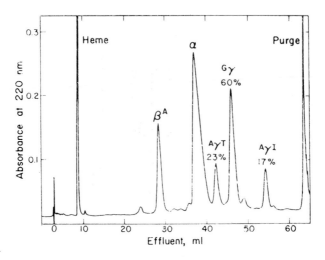

Figure 9. Separation of the globin chains in the cord blood of a newborn child on a column of reversed-phase HPLC material. Column, 25 cm × 4.6 mm; eluant, 44−56.5% of 40% aqueous TFA in 80% aqueous TFA; flow-rate, 1.0 ml/min.

will cause considerable tailing of protein and peptide peaks, resulting in lower resolution. The material (trade marked Vydac), successfully separates macromolecules from haemoglobins to small peptides, and is available as the C_{18}, C_4 or diphenyl bonded phases. *Figure 9* illustrates the separation of globin chains in the cord blood of a newborn child and is taken from some recent work carried out on a Vydac C_4 column by Shelton *et al.* (13), using a linear gradient between 0.1% aqueous TFA and 0.1% TFA in methanol. The chromatogram shows sharp, well-separated peaks which correspond to the major alpha, beta and gamma globin chains. The sharpness of the peaks also reveals the existence of minor components which were present in all the samples analysed by the Vydac column.

The technique is also effective in resolving other globin chains: the delta chain, for instance, is well-separated from the beta chain and from other beta chains such as beta(C), beta-O(Arab) and beta(E). When globin chains are isolated in this way, the amino acid analyses are in excellent agreement with the expected values: no more than 50−100 μg of an individual chain from a single chromatogram is needed for an analysis. Moreover, the solvent system used by this analysis is volatile, allowing for subsequent operations, such as amino acid analysis and sequencing, to be done after evaporation of the solvent.

One additional parameter in the chromatography of peptides is pH. Depending on the isoelectric point, the chromatography of small peptides will be affected by pH. Several small peptides (8−13 amino acids plus insulin) show one elution order when chromatographed at pH 2 and a distinctly different order when chromatographed at pH 10. This property can be utilized to optimize separation or, more importantly, to confirm that a peak in a chromatogram is in fact a single component.

A recent development, therefore, is the production of protected silica based reversed-

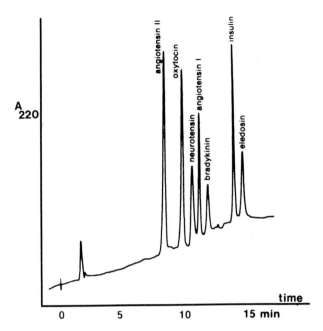

Figure 10. Separation of a mixture of peptides at pH 10.0. Column, Vydac 228TP104; gradient, $10-100\%$ B in 30 min; solvent A, 4% NH_4 acetate adjusted to pH 10 with NH_4OH; solvent B, 0.1% NH_4 acetate in 30:70 $H_2O:CH_3CN$ adjusted to pH 10 with NH_4OH; detection, UV at 220 nm.

phases, like Vydac 228TP, a C_8 bonded phase that will withstand pHs from 2 to 10 (*Figure 10*).

Tryptic digests can now be routinely separated on C_4 reversed-phase HPLC columns and it has been suggested (14) that the complex chromatograms that are produced can be better understood by re-chromatographing early eluting peaks on a C_{18} column, while late eluting fragments that may be impure may be re-chromatographed on a diphenyl column, taking advantage of the different selectivity of the aromatic phase. Column switching techniques may be employed to automate this technique.

SynChropak offer a variety of pore sizes for proteins and peptides, and they suggest that 10 nm pore size is optimal for small peptides, while 30 nm is suitable for large peptides such as those from CNBr cleavage. In addition a 100 nm pore size reversed-phase is available for very large proteins.

Shorter chain lengths, as in the spherical 5 micron, 30 nm C3 phase produced by Beckman (Altex Division, USA), provide a method of speeding up analysis times.

Conventional reversed-phase HPLC columns, however, need not be ignored for this area of application. Histones, for example (15), have been chromatographed on MicroBondapak C18, while Ultrasphere ODS (Beckman Altex) with a small pore size of $8-9$ nm is also often used.

Recently, polymeric reversed-phase materials have been developed that have the polystyrene−divinyl benzene co-polymer support, previously discussed for other

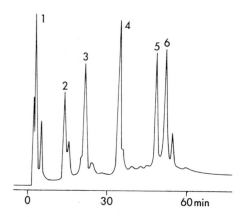

Figure 11. Separation of a protein mixture by hydrophobic interaction chromatography using a TSKgel Phenyl-5PW column. Column, 7.5 cm × 7.5 mm; eluant A, 0.1 M phosphate buffer, pH 7.0 + 1.7 M $(NH_4)_2SO_4$; eluant B, 0.1 M phosphate buffer, pH 7.0; A → B 60 min linear gradient; flow-rate, 1 ml/min; detector, UV at 280 nm. Peaks: (**1**) cytochrome c; (**2**) myoglobin; (**3**) ribonuclease; (**4**) lysozyme; (**5**) α-chymotrypsin; (**6**) α-chymotrypsinogen.

applications, but with a wide pore and C_{18}-bonded surface. Their main advantage is the facility to be able to use the complete pH range, although in general they may not be as suited to scale up as silica based supports.

2.4 Hydrophobic interaction chromatography

A further extension to reversed-phase chromatography is provided by a series of weakly hydrophobic HPLC supports that are used with milder gradient eluting conditions. Hydrophobic interaction chromatography (HIC) is a technique that complements both ion-exchange and reversed-phase HPLC and rapidly resolves biopolymers on the basis of their difference in hydrophobic interaction with the surface of the column support. A totally aqueous eluant is used, facilitating the retention of biological activity and tertiary structure.

2.4.1 Choice of HIC column

Most supports available are bonded in a similar manner to that described for reversed-phase materials. SynChropak, for example uses a 6.5 micron spherical silica which is first covalently bonded with a polyamide coating before derivatizing with a hydrophilic ligand such as propyl, pentyl or hydroxypropyl. They are compatible with buffers in the pH range 4−8. Using the same approach on their polymeric G5000PW support, the Toya Soda Company produce two types of hydrophobic interaction columns that expand the usable pH range from 2 through to 12. *Figure 11* gives an example of a protein separation on TSKgel Phenyl, while *Table 5* summarizes the range available of both wide pore silica and polymeric based supports.

2.5 Hydroxyapatite columns

A different inorganic column support material, based on microcrystalline hydroxyapatite (HA), $Ca_{10}(PO_4)_6(OH)_2$, is thought to be capable of handling much larger proteins,

Table 5. Column supports for hydrophobic interaction HPLC of proteins.

Support	Functional group	Support base	Particle size (μm)	Column size available length (cm) × i.d. (mm)	Manufacturers
Bakerbond HI-Propyl	$CH_2CH_2CH_3$	Silica	5	25 × 4.6	J.T.Baker
Poly PROPYL A	$CH_2CH_2CH_3$	Silica	5	20 × 4.6	PolyLC
Poly ETHYL A	CH_2CH_3	Silica	5	20 × 4.6	PolyLC
Poly METHYL A	CH_3	Silica	5	20 × 4.6	PolyLC
SynChropak Propyl	$(CH_2)_2CH_3$	Silica	6.5	25 × 4.6,10	SynChrom Inc.
SynChropak Hydroxy	$(CH_2)_2CH_2OH$	Silica	6.5	25 × 4.6,10	SynChrom Inc.
SynChropak Methyl	CH_3	Silica	6.5	25 × 4.6,10	SynChrom Inc.
SynChropak Pentyl	$(CH_2)_4CH_3$	Silica	6.5	25 × 4.6,10	SynChrom Inc.
TSKgel Ether-5PW	Oligo-ethyleneglycol	Polymer	10	7.5 × 7.5,15 × 21.5	Toya Soda Co.
TSKgel Phenyl-5PW	Phenyl	Polymer	10	7.5 × 7.5,15 × 21.5	Toya Soda Co.

21

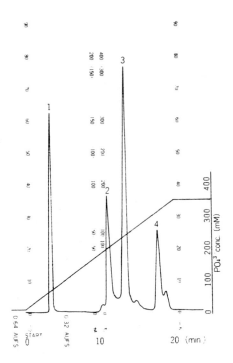

Figure 12. Separation of a protein mixture using a hydroxyapatite column (HCA-A-7610). Detector, UV at 280 nm, 0.32 AUFS; eluant, potassium phosphate buffer, pH 6.8 linear gradient elution; flow-rate, 10 ml/min; temperature, ambient; chart speed, 5 mm/min; sample, 20 μl. Peaks, (**1**) tryptophane (0.18 mg/ml); (**2**) BSA (7.5 mg/ml); (**3**) lysozyme (2.5 mg/ml); (**4**) cytochrome *c* (1.4 mg/ml).

greater than 10^4 daltons. They are generally stable over pH 4 and *Figure 12* gives an example of their application to macromolecules.

Since hydroxyapatite has Ca^{2+}, a positive charge centre, and phosphate, a negative charge centre, much can be explained by an ion-exchange absorption theory, that is that either or both of these centres are electrostatically united to a part of the protein with the opposite electronic charge. Yet it is true that there are phenomena that make it quite different from ordinary ion-exchange chromatography—such as the fact that only some ions can elute absorbed protein and the fact that what shows as a single peak by conventional ion-exchange chromatography is sometimes separated by hydroxyapatite chromatography. The true mechanism is thus yet to be fully understood.

2.5.1 *Choice of HA column*

As it is only relatively recently that techniques have been found to produce microcrystalline hydroxyapatite with a narrow particle size distribution suitable for HPLC, the selection available is still small. The more familiar ones are summarized in *Table 6*.

2.6 **Affinity chromatography**

One further technique for the separation of macromolecules that has had considerable

Table 6. Commercially available hydroxyapatite columns.

Column support	Particle size (µm)	Column sizes length (cm) × i.d. (mm)	Manufacturers
HCA-Column (A-7610)	NA	7.5 × 4,10 × 7.6	Mitsui Toatsu Chemicals Inc.
Koken hydroxyapatite	NA	10,30,60,120 × 6	Koken Co. Ltd
Pentax	2,10,20	1,5,10 × 4.0, 1,5,10 × 7.5	Pentax
TSKgel HA-1000	5	7.5 × 7.5	Toya Soda Co.

NA = not available.

Table 7. Choice of ligand for affinity chromatography.

Ligand	Examples of protein separation
Cibracron blue F3G-A	Albumin, NAD-dependent enzymes, cell growth factors, interferons, transferases, polymerases
Boronate (*m*-amino phenyl boronic acid)	Glycoprotein, transfer RNA, nucleotides, nucleosides, catecholamines
Heparin	Coagulation factors, lipoproteins, lipoprotein lipases
p-Amino benzamidine	Trypsin, bovine thrombin, urokinase, enterokinase, plasminogen activator
Imino di-acetic acid	Protein fractionation as Cu or Fe complexes
Concanavalin A	Interferon, α-antitrypsin, human alkaline phosphatase
Dextran blue	Certain kinases, dehydrogenases
Adenosine monophosphate	Dehydrogenases
Adenosine triphosphate	Protein kinases

success in recent years is affinity chromatography. Instead of relying on differences in size, polarity, electronic charge or other physical and chemical differences, affinity chromatography bases its mode of separation on a unique, reversible and highly selective interaction between the macromolecules and the surface, similar to that of an enzyme or co-enzyme, using a lock and key type mechanism. Immobilizing a specific ligand onto the surface of a stationary phase allows the separation of a wide variety of large proteins, glycoproteins and enzymes, the ligand being carefully chosen to target the protein of interest. One example is the use of co-enzyme B bonded agarose to separate and purify the vitamin B-binding proteins, transcobalamin 1 and 11 and methylmalonyl-CoA mutase. Agarose-bonded lectins have been used to purify various glycoproteins, while immobilized drugs, vitamins, peptides and hormones (16) have all been used to isolate corresponding receptors or transport proteins. Conversely, of course, lectins can be separated by glycoprotein bonded onto agarose.

In practice, a specific ligand is covalently attached to a solid support, the material packed into a column and the enzyme or protein in a crude mixture injected. The column is washed to remove unabsorbed, contaminating proteins and the proteins of interest eluted by changing pH, ionic strength or by adding a stronger ligand to the mobile phase (17). The usual solid supports are polysaccharide based, in particular beaded agarose (sometimes strengthened by crosslinking with epichlorohydrin), although polyamides and microporous glass beads have also been used. As with other HPLC techniques, the method of immobilizing the ligand is crucial, and much has been published in the literature on their preparation (18), which usually involves the inclusion of a spacer arm, such as hexamethyldiamine, between the ligand and support. Again, like other HPLC techniques, the support needs to be mechanically strong and sufficiently hydrophilic to avoid non-specific binding of proteins, while the spacer arm must be carefully chosen so as not to introduce any unnecessary hydrophobicity.

2.6.1 *Choice of affinity column*

Whilst the preparation of agarose based phases for affinity chromatography is often carried out by the chromatographer, because the choice of ligand is so specific to each application, a vast range of affinity media is also commercially available on agarose, crosslinked agarose, dextran and polyacrylamide beads (for example, Pierce Chemical Company). However, the use of such soft gels has the same drawbacks for affinity chromatography as had been found in the early days of reversed-phase HPLC, namely, restricted flow-rates and therefore slow separation times. In order to retain biological activity during these long time periods, it is often necessary to carry out the procedure at low temperature. More importantly, much depends on the ease and reproducibility that the chromatographer can achieve in binding the ligand to the support surface in the first place. Problems such as leaching of the ligand following binding are common. Fortunately, they are beginning to be superceded by high-performance silica and polymer based supports, and in fact the discussion here will be restricted to these true HPLC materials since the other types are discussed in Chapter 6.

The choice of ligand to facilitate a particular separation is wide and *Table 7* gives examples of those used. The wide range listed includes triazine dyes, some of which are particularly good for the separation of serum albumin (19). In summary, the potential ligand for affinity chromatography must satisfy the following requirements.

(i)   The ligand must exhibit a specific and reversible binding affinity for the compound of interest.
(ii)   The ligand must possess appropriate functional groups which allow immobilization without compromising the activity of the ligand.
(iii)   The interaction between the ligand and the compound of interest must be sufficient but not so strong that elution is not readily achieved.

Having chosen a suitable ligand, there are two alternative approaches in the choice of column. Firstly, many columns are now available with the ligand required already bound to silica or to polystyrene−divinyl benzene co-polymer (*Table 8i*). However, if the chosen phase is not available commercially, preparation can often be readily achieved via one of the recently developed supports that are bonded with reactive functional groups. Ultraffinity-EP (Beckman Instruments Inc., Berkeley, CA), features an active epoxide function bonded to macroporous silica via a hydrophilic spacer. The epoxide group will react readily with ligands having primary amino, sulphydryl and hydroxyl moieties to form stable covalently-bonded linkages. Another approach used by the Pierce Company is to employ the trifluoro-derivative shown below, which will

Table 8. (i) Commercially available columns for affinity HPLC of proteins.

Column support	Bonded ligand	Support	Column size length (cm) × i.d. (mm)	Manufacturers
ABA - 5PW	p-Amino benzamidine	Polymer	7.5 × 7.5,15 × 21.5	Toya Soda Co.
Blue - 5PW	Cibracron blue F3G-A	Polymer	7.5 × 7.5,15 × 21.5	Toya Soda Co.
Boronate - 5PW	m-Aminophenylboronic acid	Polymer	7.5 × 7.5,15 × 21.5	Toya Soda Co.
Chelate - 5PW	Imino di-acetic acid	Polymer	7.5 × 7.5,15 × 21.5	Toya Soda Co.
Heparin - 5PW	Heparin	Polymer	7.5 × 7.5,15 × 21.5	Toya Soda Co.
Selecti Spher - boronate	Boronic acid	Silica	5 × 5,10 × 5,25 × 5	Pierce
Selecti Spher - Cibracron blue	Cibracron blue F3G-A	Silica	5 × 5,10 × 5,25 × 5	Pierce
Selecti Spher - Concanavalin A	Concanavalin A	Silica	5 × 5,10 × 5,25 × 5	Pierce

(ii) Commercially available reactive HPLC columns for affinity chromatography of proteins.

Trade name	Active surface grouping[a]	Support	Column size length (cm) × i.d. (mm)	Manufacturers
Affi-prep-10	N-hydroxysuccinimide	Polymer	3 × 4.6,1.5 × 25	Bio Rad
Affin-matrix Diazofluoborate	$R-N_2^+$	Silica	Bulk material ony	J.T.Baker
Affin-matrix Epoxypropyl	Epoxide	Silica	Bulk material only	J.T.Baker
Affin-matrix Glutaraldehyde	$-C=C-CHO$	Silica	Bulk material only	J.T.Baker
Selecti Spher-Activated Tresyl	$-O-CH_2-CH(OH)-CH_2-$ $OSO_2-CH_2-CF_3$	Silica	5 × 5,10 × 5,25 × 5	Pierce
Glycidoxypropyl	Epoxide	Silica	Bulk material only	J.T.Baker
Affin-matrix p-Nitrophenyl	$\overset{O}{\underset{\parallel}{-C}}-O-C_6H_4-NO_2$	Silica	Bulk material only	J.T.Baker
Ultraffinity-EP	Epoxide	Silica	5 × 4.6,10 × 10	Beckman Inc.

[a]All except glycidoxypropyl have spacer arms (hydrophilic).

25

Table 9. Commercially available columns for HPLC of immunoglobulins.

Column support	Column type	Bonded ligand	Support	Column sizes length (cm) × i.d. (mm)	Manufacturer
Accell QMA	Anion-exchange	-N$^+$	Silica	Bulk material	Waters Millipore
Affi-Prep Protein A	Affinity	Protein A	Polymer	3 × 4.6 and in bulk	Bio-Rad Laboratories
Bakerbond ABx	Mixed mode	Weak anion/cation	Silica	10 × 7.75	J.T.Baker
Bakerbond MAb	Anion-exchange	'Weak anion'	Silica	10 × 7.75	J.T.Baker
BioRad HPHT	Mixed mode	Hydroxyapatite	Silica	10 × 7.8,30 × 4.6	Bio-Rad Laboratories
Glutaraldehyde	Reactive affinity	-C=C-CH=O	Silica	Bulk material only	J.T.Baker
Mono Q	Anion-exchange	Quaternary amine	Polymer	5 × 5	Pharmacia
PL-AFc	Affinity	Protein	Polymer	5,15 × 4.6,5 × 7.5	Polymer Laboratories
SelectiSpher-10	Affinity	Protein G	Silica	5 × 3.5 × 5	Perstorp Biolytica
SelectiSpher-Protein A	Affinity	Protein A	Silica	5 × 3.5 × 5	Perstorp Biolytica
TSK DEAE	Anion-exchange	DEAE	Polymer	7.5 × 7.5	Toya Soda Co.
Ultraffinity-EP	Reactive affinity	Epoxide	Silica	5 × 4.6,100 × 0.10	Beckman Inc.
Zorbax GF250	Gel filtration	–	Silica	2.5 × 9.4	Du Pont

immobilize a ligand if it contains an available primary amino or thiol group:

$$Si\text{-}O\text{-}Si\text{-}(CH_2)_3\text{-}O\text{-}CH_2\text{-}CH(OH)\text{-}CH_2\text{-}OSO_2CH_2CF_3 + (Ligand)\text{-}NH_2$$
$$\rightarrow -Si\text{-}O\text{-}Si\text{-}(CH_2)_3\text{-}O\text{-}CH_2\text{-}CH(OH)\text{-}CH_2\text{-}NH\text{-}(Ligand)$$

3. COLUMNS AND PACKING MATERIALS FOR IMMUNOGLOBULINS

A rapidly growing number of applications in the diagnostic and therapeutic field requires large quantities of high purity antibodies and this has led to remarkably fast development in the area of antibody separations. Moreover, the techniques available have been further advanced so that they can readily be used in a preparative mode. The initial aim of these separation studies was not only to be able to separate IgG, the immunoglobulin of principal interest, from other interfering proteins, but also to be able to separate the other four classes of immunoglobulins, and their sub-classes. In addition, as the current trend is for the production of antibodies by cell culture methods, rather than isolation from serum, the need for more sophisticated HPLC methods is more apparent since cells secrete multiple types of antibodies including monoclonal and myeloma immunoglobulins.

3.1 **Choice of column for immunoglobulins**

Two main approaches to this separation problem have been used, the first utilizes affinity chromatography in which highly specific biosorption of the immunoglobulin of interest occurs. Indeed much of the discussion in Section 2.6 on affinity chromatography of proteins also applies here, although there are additional column support materials available that are applicable to immunoglobulins only (see *Table 9*). Protein G, for example, is a bacterial cell wall protein from strains of human group G streptococci, and binds strongly to the Fc region of IgG (20). It will also bind to all IgG sub-classes to a certain extent, while protein A generally does not show sufficient affinity for these substances (21). Both these proteins are available immobilized by covalent bonding on silica or polymeric supports for use in HPLC. The use of reactive affinity supports, such as Ultraffinity-EP (Beckman Altex) and Affi-Prep 10 (Pierce Chemical Co.), whereby the active sites of the HPLC column are derivatized with a targeting species, such as protein A, is also common.

The second method utilizes a traditional method for antibody purification, that of ion-exchange. The particular functional moiety varies somewhat, ranging from weak anion-exchange to quaternary amino groups that are quite strongly basic. The general mechanism of separation seems to be similar regardless of functional group, however, and binding of the three major ascites proteins, transferrin, immunoglobulin and albumin occurs in that order, with maximum resolution being achieved by optimizing gradients of ionic strength and pH. One weak anion-exchanger, Bakerbond MAb (described as having a broad range of pK_as that are weakly basic) has been shown to be successful in many IgG separations, resembling a DEAE-bonded phase in its specificity. A typical chromatogram is shown in *Figure 13*. Other weak anion-exchangers, like the polymeric TSK DEAE, generally give shorter retention times, but sometimes with overlap of albumin and IgG peaks. The polymer beads used for Mono Q have physical properties similar to TSK DEAE, but with a limitation of 2 ml/min flow-rate. By bonding

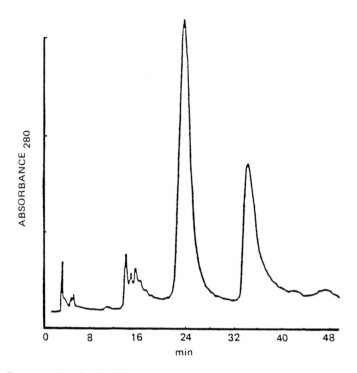

Figure 13. Chromatography of ascites fluid on Bakerbond MAb HPLC columns; capacity for IgG purification. Eluant A, 10 mM KH_2PO_4, pH 6.81; eluant B, 0.5 M KH_2PO_4, pH 6.40; gradient $0-25\%$ B in 60 min; 4.6×250 mm MAb column 15 μl ascites fluid.

anion-exchanger onto silica, as in Accell QMA (Waters Millipore), easier scale-up is possible, though with lower resolution because of particle size.

The utilization of column support materials that resolve immunoglobulins by multiple mechanisms have also become commonplace in this area of HPLC. The mixed mode ion-exchange material in Bakerbond ABx (J.T.Baker) has been used to purify antibody from ascites fluid, serum-based tissue culture and plasma, as exemplified in *Figure 14*.

Another approach in the mixed mechanism type of column is the use of hydroxyapatite, thought to have an advantage over ion-exchange in that because the antibody elutes at a higher phosphate concentration than does the bulk of contaminating proteins, it is possible to choose a buffer concentration such that the interfering proteins pass directly through the column, leaving the bulk of binding capacity for immunoglobulins. Hydroxyapatite is a crystalline form of calcium phosphate, traditionally used for open column chromatography, but used in microparticulate form for HPLC.

4. COLUMNS AND PACKING MATERIALS FOR OLIGONUCLEOTIDES

Increased research in DNA structure and gene cloning has led to a proliferation of machines for the synthesis of oligonucleotides. This, in turn, has created a need for more rapid and higher resolution purification of fragments and oligomers of DNA and RNA. New columns and techniques now provide versatile and efficient methods for

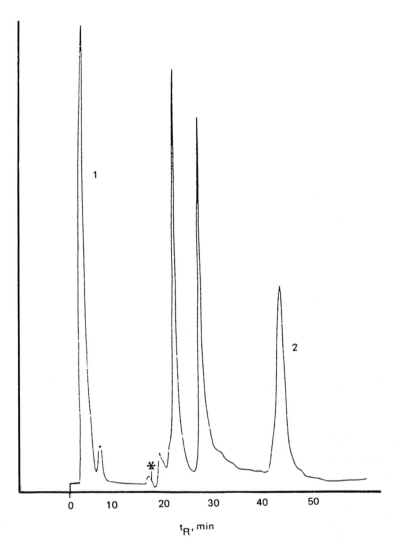

Figure 14. Selective binding of mouse ascites IgG to Bakerbond ABx: purification to over 99%. A 25 µl sample of mouse ascites was diluted 4-fold in initial buffer and chromatographed on a Bakerbond ABx analytical column (4.6 × 250 mm) with a 50 min linear gradient of 10 mM KH_2PO_4, pH 6.0, to 125 mM KH_2PO_4, pH 6.8. **Peak 1** (the void volume peak) contains albumin, transferrin, and other proteins totalling over 95% of non-IgG protein in ascites fluid. **Peak 2** is the IgG peak. Asterisk denotes 4-fold increase in detector sensitivity.

the analysis of both high molecular weight nucleic acids and their constituent bases (*Table 10*).

Hydrolysis of DNA can be accomplished in a variety of ways, using enzymes, acids or bases, giving rise to many different derivatives of the purine and pyrimidine units. The actual hydrolysis method used will determine the chromatographic technique employed. The neutral hydrolysis condition used with enzymes is considered to yield just four deoxynucleoside triphosphates but there may also be present, however, small

Table 10. Columns available commercially for the separation of oligonucleotides and related products.

Column support	Type of support	Bonding chemistry	Species separated	Pore size (nm)	Column sizes (μm)	Bulk material available	Manufacturers
Partisil 10SCX	Strong cation-exchanger	Sulphonic acid	Purine bases		250 × 4.6	Yes	Whatman
Nucleosil 10SA	Strong cation-exchanger	Sulphonic acid	Purine bases	10	250 × 4.6	Yes	Macherey Nagel
Vydac 214TP54	Reversed-phase	C$_4$	Oligonucleotides	30	250,150 × 4.6	No	The Separations Group
Vydac 3040L	Weak anion-exchanger	DEAE	Oligonucleotides	39.9	250,150 × 4.6	No	The Separations Group
Nucleogen 60-7	Weak anion-exchanger	DEAE	Oligonucleotides	6	125 × 4,10	No	Macherey Nagel
Nucleogen 500-7	Weak anion-exchanger	DEAE	RNA	50	125 × 6,10	No	Macherey Nagel
Nucleogen 4000-7	Weak anion-exchanger	DEAE	Plasmids, DNA, RNA	400	125 × 6,10	No	Macherey Nagel
SynChropak AX100	Weak anion-exchanger	Polymer	Oligonucleotides	100	250,100 × 4.6,10	Yes	SynChrom Inc.

quantities of derivatives—5'-methyl dCTP being of particular interest. Further hydrolysis of the triphosphates gives diphosphates, monophosphates, deoxynucleosides and finally the purine and pyrimidine bases. Selectivity can be achieved by using 0.1 M acid at pH 1 and 70°C to give just the purine, guanine and adenine bases and their alkylated derivatives. Hydrolysis with 40% HF gives both purines and pyrimidines, and the acid is removed by evaporation. For molecular biology laboratories, HPLC column supports need to be able to separate some or all of these products, in addition to large DNA molecules themselves.

4.1 **Choice of oligonucleotide column**

HPLC of the purines can be successfully accomplished on the strong cation-exchangers, Nucleosil SA or Partisil SCX, giving good resolution with an eluant of $5-50$ mM sodium acetate at pH 4. The molarity used will depend on the age of the column, a lower one being needed as the lifetime of the column increases. Relative retention times of adenine and its derivatives can be altered by the addition to the buffer of modifiers such as methanol, thus providing a convenient way of tailoring the chromatography to suit the needs of the user. Temperature programming is also useful to increase the speed of analysis, especially when 3'-methyl adenine is of interest, as retention times in excess of 90 min at 25°C are common, but decrease to 45 min at 60°C.

HPLC assay of nucleoside mono-, di- and triphosphates is normally achieved with a phosphate gradient on a strong cation-exchanger. A typical gradient would be from 5 mM to 1 M KH_2PO_4, pH $2.85-4.45$. If only one group is of interest then isocratic conditions extrapolated from the gradient can be used.

Reversed-phase columns have also been used for analysing most of the hydrolysis products of RNA and DNA and these have two advantages over ion-exchange in being the more commonly available columns, and in giving a quicker and sharper separation. It is necessary in this case to suppress the formation of highly charged ions, such as occurs in nucleotide chemistry and this can be achieved by either using a high molarity phosphate buffer (22), choosing a pH lower than the pK_a, or by ion-pairing methods (23). For reversed-phase HPLC of the bases, a typical buffer would be 50 mM ammonium phosphate at pH 5.9 (22) . A new development is the use of well protected C_4 phases, such as Vydac 214TP (The Separations Group) to enable the separation to occur without the interference of residual silanol groups. Using these columns, fast and efficient purification of oligodeoxyribonucleotides up to 104 nucleotides in length has been achieved (24).

The synthesis of oligonucleotides yields a product that is a mixture of tritylated and non-tritylated species. Analysis of the product is generally carried out again on Vydac C_4 (25), a column support that is able to separate not only the two classes, but also allows some separation of the various tritylated species.

Larger fragments can now also be chromatographed using a weak ion-exchange HPLC column. Bonding DEAE moieties onto silica as in Nucleogen DEAE-60-7 (Macherey Nagel), or in Vydac 3040L (The Separations Group), provides unique selectivity for the purification of oligonucleotides up to 68 residues long (*Figure 15*). Preliminary work on Vydac 3040L has indicated that even larger fragments up to 75 residues may be purified by the adjustment of eluant conditions, using ammonium sulphate as the

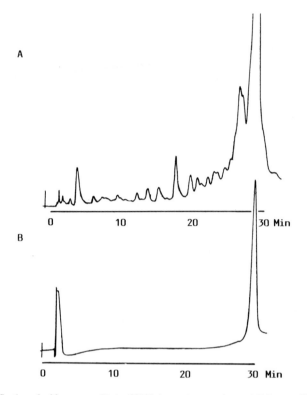

Figure 15. Purification of a 20-mer on a Vydac 3040L ion-exchange column. (**A**) Impure 20-mer. (**B**) Purified 20-mer.

eluting salt. The wide pore Nucleogen DEAE 4000-7 can also isolate plasmids, ribosomal RNA, messenger RNA and viral RNA (*Figure 16*). A remarkable success of this column is the isolation of supercoiled peptide DNA from a crude cell lysate in a single step in under 40 min (26).

Two other silica based weak anion-exchangers, SynChropac AX100 (SynChrom Inc.) and Polyanion SI (Pharmacia) will also separate oligonucleotides and nucleotides, in the latter case up to 14 bases long.

5. COLUMNS AND PACKING MATERIALS FOR OLIGOSACCHARIDES

Historically, oligosaccharide analysis has been performed by GLC, thin-layer chromatography and paper chromatography, methods that require long analysis times and possible derivatization steps. There is now a vast range of carbohydrate columns available for HPLC, making it the technique of choice and these include ion-exchange, amino-bonded silica and cation-exchange polymers.

5.1 Choice of oligosaccharide column

Most carbohydrate applications require a column that is capable of analysing a mixture of sugar molecules for their unit length and for their individual epimer forms. The first

32

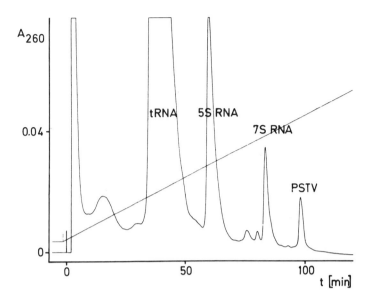

Figure 16. Separation of RNAs from plant extracts. Sample, viroid (PSTV) infected plant RNA extract; column, Nucleogen®-DEAE 500-7; size, 6 mm i.d. × 125 mm; eluant A, 250 mM KCl, 20 mM phosphate buffer, pH 6.6, 5 M urea; eluant B, 1 M KCl, 20 mM phosphate buffer, pH 6.6, 5 M urea; gradient, 100% A (0% B)−100% B (0% A), linear; flow-rate, 1 ml/min, 36 bar; temperature, ambient.

HPLC columns to be used were amino-bonded silica gels that used an eluant of acetonitrile and water (27). Whilst this method gives a shorter retention time for disaccharides than earlier ion-exchange methods, resolution is poor by comparison. Additionally, the high percentage of organic solvents required (generally >75%) makes analysis expensive and, more importantly, on-line analysis of carbohydrates in food related industries hazardous. For larger oligosaccharides, interaction with the amino surface can be too strong, giving long analysis times. Other methods include the use of reversed-phase columns using water as the mobile phase (28), or using ion-pairing reagents (29) but both give insufficient resolution generally for oligosaccharides.

The use of cation-exchange polymeric column supports has gained considerable attention due to its simplicity: most use 100% water as the mobile phase. Most are based on crosslinked polystyrene−divinyl benzene co-polymers and separate principally by steric exclusion (enabling distinction between the saccharide chain lengths), but ligand formation and partitioning effects may play some role resulting in separation of a number of compounds with similar or identical molecular weights. However, because of the complexity of this class of macromolecule, no simple HPLC column can be optimized to separate the complete range of oligosaccharides. Different selectivities are observed depending on the ionic form of the ion-exchange polymer and, commercially, columns are provided with calcium, sodium, potassium, lead and silver counter-ions, to cover a range of different applications (see *Table 11*). For example, a column containing cation-exchange resin in the calcium form is able to resolve the oligomers Dp3 and Dp4, plus the monosaccharides glucose, galactose and fructose (*Figure 17*).

Table 11. Commercially available cation-exchange polymeric columns for oligosaccharides.

Column support	% Cross-linking	Ionic form	Particle size (µm)	Column size length (cm) × i.d. (mm)	Examples of separations	Manufacturers
Aminex HPX-87N	8	Sodium	9	30 × 7.8	Raffinose, sucrose, fructose	Bio-Rad
Aminex HPX-87K	8	Potassium	9	20 × 7.8	Glucose, maltose, maltotriose, Dp4	Bio-Rad
Aminex HPX-42C	4	Calcium	25	30 × 7.8	Dp4, 3, 2, xylose, arabinose, mannose Dp14	Bio-Rad
Aminex HPX-42A	4	Silver	25	30 × 7.8	Dp14 to Dp2	Bio-Rad
Aminex HPX-65A	6	Silver	11	30 × 7.8	Dp6 to Dp3	Bio-Rad
Bio Rad Fast Carbohydrate	8	Lead	9	30 × 7.8	Maltose, fructose	Bio-Rad
Interaction CHO-411	NA	Sodium	20	30 × 7.8	Dp15 to Dp2	Interaction Chemicals
Interaction CHO-611	NA	Sodium	12	30 × 6.5	Dp4 to Dp2	Interaction Chemicals
Interaction CHO-682	NA	Lead	8	30 × 7.8	Disaccharides	Interaction Chemicals
MC1 Gel CK02A	2	NA	20–24	Bulk material only	Oligosaccharides in general	Mitsubishi
MC1 Gel CK04S	4	NA	11–14	Bulk material only	Oligosaccharides in general	Mitsubishi
MC1 Gel CK04C	4	NA	17–20	Bulk material only	Oligosaccharides in general	Mitsubishi
MC1 Gel CK06S	6	Sodium	11–14	Bulk material only	Oligosaccharides in general	Mitsubishi
Polypore CA	NA	Calcium	10	22,10 × 4.6	Maltotriose and higher oligomers	Brownlee
Shodex ION pak KS-801	NA	NA		50,25 × 8	Dp8 to Dp1	Showa Denko
Shodex ION pak KS-802	NA	NA		50,25 × 8	Dp10 to Dp1	Showa Denko
Shodex ION pak KS-803	NA	NA		50,25 × 8	Polysaccharides	Showa Denko
Shodex RS pak DC-613	NA	Sodium		15 × 6	Mono-, di-, tri- and oligosaccharides	Showa Denko
Shodex SUGAR SH 1011	NA	Hydrogen		30 × 8	Oligosaccharides in general	Showa Denko
Spherogel Carbohydrate	NA	Calcium	10	30 × 4.6	Galactose, mannitol, sorbitol	Beckman Insts
TSKgel G-Oligo-PW	NA	NA		30 × 7.8	Oligosaccharides in general	Toya Soda Co.

NA = not available.

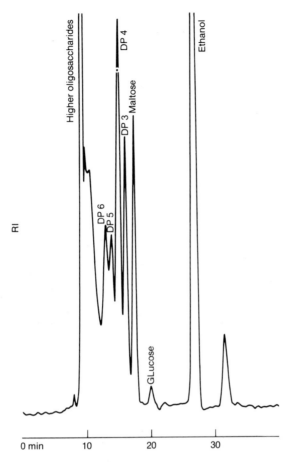

Figure 17. Sugar profile of domestic beer. Several oligosaccharides, along with monosaccharides and ethanol, are resolved in this domestic beer sample using a lead bonded polymeric column. Column, CHO-682 (Interaction Chemicals); eluant, water; flow-rate, 0.4 ml/min; temperature, 80°C; detection, refractive index (RI).

6. COLUMN CARE AND SUPPLY

6.1 Column packing techniques

Microparticulate silica based packing materials may be packed into columns in-house using now well-developed slurry packing procedures. Owing to the much narrower particle size distribution of current materials, sizing prior to packing is no longer necessary, and the use of balanced density slurry methods, involving noxious solvents such as tetrabromoethane, have been virtually abandoned. Solvents still have to be carefully chosen to minimize particle aggregation, and commercially it is usual for the slurry and packing solvents to be optimized for each particular phase. Additionally, polar adsorbents such as unbonded silica gel will not pack well in identical solvents chosen for the very high hydrophobic octadecylsilane bonded phases.

Although much of the mystery surrounding column packing has been dispelled, beginners to the technique will find that it takes time to obtain consistently reproducible

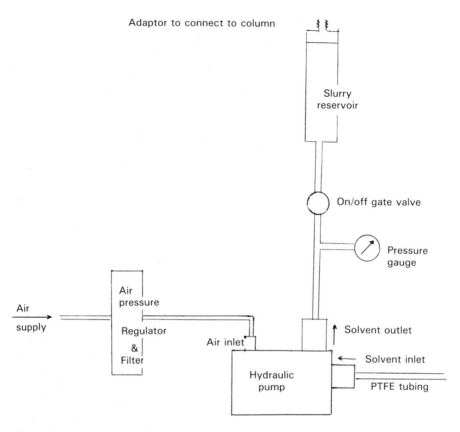

Figure 18. Typical column packing high pressure, high flow-rate generating system.

and stable columns. Several methods have been published in the literature (30−32) but unfortunately many techniques are proprietary and therefore retained by manufacturers.

For successful column packing, high solvent velocities are required. Analytical HPLC pumps are unsuitable as they do not usually have a high enough flow-rate or pressure capability. High capacity, pneumatic amplifier, constant pressure pumps are better as they provide sufficient flow-rates and pressure for analytical and semi-preparative columns. Several manufacturers market purpose built slurry packers (for example Shandon Southern Products, Runcorn, UK), or one can be assembled (see *Figure 18*) using a pneumatically driven hydraulic pump, such as a Haskel MCP 110 or MCP 71 (Haskel Energy Systems, Sunderland, UK). These are capable of flow-rates up to 300 ml/min at 8800 p.s.i., given an inlet pressure of 100 p.s.i. from a compressor supplying clean, dry air, although the maximum pressure ratings are a good deal higher than this.

To pack a column, the packing material as a sonicated slurry is poured into the slurry reservoir of the packing rig, the column connected and the pressure applied. Calculation of the correct quantity of packing material is carried out by multiplying the volume of the column tube by the packing density: most silica manufacturers will state the packing density of their material in their literature (typically 0.7 g/ml) and some will also suggest suitable packing procedures. Typical solvents for column packing are methanol or dry acetone for both slurry and packing media, and usually about 300−400 ml of solvent is pumped through a 25 cm long (4.6 mm i.d.) column during the packing procedure. After the column is packed, turn off the air supply but leave the column connected to the packing rig for about 5 min to allow the pressure within the column to return to normal. The column is then disconnected and a small amount of excess slurry applied with a microspatula to the top of the column so that the column bed stands slightly proud. Once the column end fitting and frit are connected, the column can be tested.

It is generally far more difficult to pack polymeric based materials because of the need to avoid over-compressing the bed. The choice of solvent and packing pressure is consequently far more critical. For this reason, many of these materials are only available in pre-packed columns.

6.2 **Pre-packed columns**

For chromatographers who do not use a sufficient number of columns per year to economically justify the purchase of a HPLC column packing facility, or who cannot afford the time sometimes required to obtain consistent results, the purchase of pre-packed columns is advised. Indeed, because some materials are difficult to pack, many manufacturers do not release the bulk material (see for example *Tables 2* and *4*). The quality and reliability of pre-packed columns has improved considerably over the last few years, and most manufacturers now provide a test chromatogram in which a standard procedure is used to calculate column performance: often a written guarantee is supplied. The test report will specify the efficiency of the column expressed in terms of the number of theoretical plates per column or metre, and will also show the test conditions used to obtain the result (mobile phase composition, flow-rate, injection volume, detection wavelength and attenuation, temperature and test sample components). It will also indicate the back-pressure the column will typically exhibit, under the test conditions. This facility is extremely useful to the chromatographer who can, on receipt of the column, repeat the test procedure for the column. Any discrepancy in the result can be due to one or more of the reasons listed below.

(i)   Damage of column in transit, or column bed settling.
(ii)   Extra-column effects occurring in the user's instrumentation compared with the manufacturer's system. Examples are slow detector time constants, long lengths of connection tubing between the column outlet and the detector flow cell, wide bore (>0.010 inch i.d.) tubing connecting the column and injector, all of which can add artificial peak broadening.
(iii)   Different methods of measuring column efficiency (there is a choice of measuring peak width at either half-height or baseline, see Section 6.3).
(iv)   Too large a sample size.
(v)   Poor equilibration of eluant before testing (unusual with reversed-phase materials).

A manufacturer will usually replace any column which is damaged in transit, or which

Table 12. Suggested standard test procedures for monitoring column performance.

Column type	Test components	Test conditions
Reversed-phase silica, C_{18} (ODS) to C_4	Benzamide, benzophenone, biphenyl	Acetonitrile:water, 70:30 or Methanol:water, 70:30
Reversed-phase silica, C_3 to C_1 and phenyl	Benzamide, benzophenone, biphenyl	Methanol:water, 60:40
Polar bonded phases e.s. (NH_2, CN, DIOL) and unbonded silicas	Napthalene, phenyl benzoate	Chloroform:hexane, 6:94
Anion-exchange	Cytosine, adenosine and uridine monophosphates	0.02 M KH_2PO_4, pH 3.5
Cation-exchange	Urasil, guanine	0.02 M $NH_4H_2PO_4$, pH 3.5

Components of the test mixture are listed in order of increasing retention time.

has suffered bed settling. It is useful to file both the original test report and the laboratory copy and then to repeat the procedure from time to time to monitor column performance.

6.3 Test procedures

It is common for the test mixture used by a manufacturer to check the performance and quality of an HPLC column to be totally unrelated to the assay for which the column was purchased because they are made up of quite simple compounds, which are readily available in the laboratory. When setting up a test procedure for monitoring column performance over its lifetime, the same procedure as the manufacturer should be used if possible. It is useful to note that a different result for the efficiency of a column will occur for each different test method, and it is therefore important for each HPLC laboratory to standarize on one only. Suggested test mixtures, along with their respective test conditions, are outlined in *Table 12*. Commonly, two or three components are used, one with a very short retention time [typically with a k' of about 0.2, where k' is the column capacity factor $(t_0 - t_r)/t_r$ (see *Figure 19*) to assess band broadening caused by the injector and detector. Breaks in the column bed and peak tailing may not be seen in a peak with a short residence time. A second peak would have a moderate retention (k' of $1-3$), while a third should be well-retained (k' of $5-8$) as a better test of the column performance alone, as instrumentation deficiencies will have a smaller effect on a peak having a long retention time.

From the test chromatogram, the following measurements are made for each peak (see *Figure 19*).

(i)   Retention time, t_r, from the marker point of injection, measured in mm.

(ii)   Peak width, w, (in mm) at half-height (measure the total peak height from the baseline, halve the value, and mark the point half way up the peak from the baseline where the peak width should be measured).

(iii) Asymmetry (draw tangents along the upward and downward slopes of the peak and measure the distance at the baseline between the tangent and actual slope of the peak, A_a and A_b in *Figure 19*).

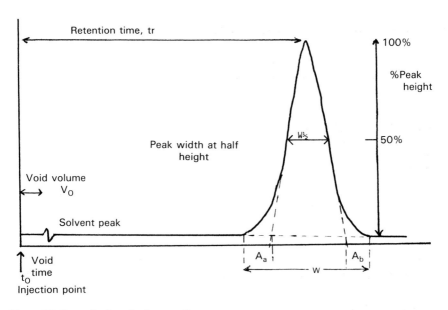

Figure 19. Determination of column performance.

The first two measurements are used to calculate the efficiency of the column:

$$N = 5.54 \ (t_r/w_{1/2})^2 \text{ plates/column}$$

The asymmetry measurements are used as a guide of deviation of the peak from its ideal Gaussian shape (see *Figure 19*):

$$A_s = A_a/A_b$$

An alternative method of calculating efficiency uses the baseline peak width (*w* in *Figure 19*) to calculate efficiency. Whilst both results should agree, if a peak is tailing (as in some bonded phases, and in ageing columns), the assignment of peak width can be difficult. For this method column efficiency is calculated using the formula:

$$N = 16 \ (t_r/w)^2 \text{ plates/column}$$

It is useful to note also that column efficiencies calculated by computer may be different from the results obtained by manual methods: the most accurate results are obtained from integrators and computers utilizing voltage-to-frequency, rather than analogue-to-digital, converters, and with a sampling rate greater than 20 points (for lower sampling rates a normalization procedure should be used).

Monitoring the column efficiency in this way will indicate how the column is lasting. Although it is difficult to state limits below which the column should be replaced, much will depend on the resolution of the sample components since for closely eluting peaks, small changes in N will affect the separation considerably.

More importantly, by routine monitoring of peak shapes and comparison with the original test chromatogram, the test procedure can be used to check suspected problems. Key areas to watch are as follows.

(i)   Band broadening. Loss of sharpness of the top of the peak occurs as the column gets older, or needs regenerating. Well-rounded or flat-topped peaks are usually a symptom of detector overload.

(ii)   Excessive peak tailing. In a reversed-phase column, this usually indicates a poorly packed bed, or interference from residual silanols, perhaps exposed by changes in the surface due to working below the minimum pH of 2.0.

(iii) Peak splitting. If this occurs on every peak, it often indicates that the column bed has dropped (leaving a void at the top of the column), or that there is a blockage in the injector (causing the sample to be injected in two segments).

(iv)   Increased back-pressure. This may be due to a simple blockage of the column inlet frit with impurities or column bed material, or to column bed damage or settling, or to operating the column over the maximum working pH of $7-7.5$. It is important to check that the increased back-pressure is in fact originating in the column: observe that the pressure reading drops to almost zero when the column is disconnected while the solvent flow is still on. A high pressure reading could also indicate a blockage in the injector or in filters associated with the pumping system.

6.4 Column regeneration

When certain problems arise (see below), then unless the column is nearing the end of its life, or has been misused, there are often regeneration or repair techniques that can be carried out.

(i)   Regeneration. If a reversed-phase, silica-based column is exhibiting band broadening or tailing, the fault may be due to retained solutes either at the top or throughout the column. Check with the manufacturer that the column flow can be reversed without damaging the column bed and, if possible, reconnect the pump to the column outlet, leaving the column inlet free; do not connect to the detector. In this way a series of solvents can be pumped through the column so that any deposited material at the top of the column may be flushed out. If the contaminant is known, the eluant can be chosen to enhance its solubility and ease its removal. If not, reversed-phase silica based columns may be regenerated by pumping the following series of variable polarity eluants.
    (a) About 10 column volumes of water, followed by
    (b) five column volumes of methanol;
    (c) five column volumes of isopropanol or dichloromethane;
    (d) five column volumes of hexane;
    (e) five column volumes of isopropanol or dichloromethane;
    (f) five column volumes of methanol;
    (g) five column volumes of water.
Use HPLC grade organic solvents and filtered, distilled water. Ion-exchange

columns may be regenerated by flushing with plenty of water, or with a buffer gradient, as some problems are caused by buffer salting out in the pores of the silica when the column has been stored in buffer rather than water (a better storage solvent still is 0.01% w/v sodium azide solution), or has had organic solvents pumped through the column without prior washing with water. For all other columns, and especially polymeric-based columns, follow the manufacturer's instructions for regeneration.

(ii)   Repair. If a column has voided, remove the column end fitting, and check the top of the column bed. Any discoloured material should be removed with a microspatula and the bed underneath flattened. The void can be filled with dry plain glass beads (40 μm), or with the same material as the column, if the material is available loose or in a top-up kit (apply in a thick methanolic paste with a microspatula). At the same time remove the frit or mesh from inside the column end fitting and clean by sonication in methanol or dichloromethane: if the frit is an integral part of the end fitting, sonicate the complete end fitting. This will often remove any blockages causing back-pressure increases.

6.5 Guard (pre-) columns

Because most HPLC columns are expensive, it is important to take as many steps as possible to increase their lifetime. Column protection devices range from simple on-line sintered stainless steel filters to elaborate low volume guard/analytical combinations. By far the most popular protection device is a guard column placed between the injector and analytical column. This can be packed with either identical material as the main analytical column, or with pellicular material of the same type. Pellicular materials are coated non-porous glass beads originally developed in the early years of HPLC but now largely superceded by porous silica materials. They have now found a place as guard column materials because they can be dry-packed, making repair and topping-up easy, and because they are available in a range of coatings to match unbonded silica, reversed-phase, polar-bonded phase and ion-exchange columns. Maintenance of all guard columns is similar, namely, check the inside of the guard column at the inlet end regularly for debris, replace any discoloured material with fresh packing and replace or clean the inlet frit if necessary.

7. REFERENCES

1.   Kato,Y., Komiya,K., Sasaki,H. and Hashimoto,T. (1980) *J. Chromatogr.*, **193**, 311.
2.   Goheen,S.C. and Matson,R.S. (1985) *J. Chromatogr.*, **326**, 235.
3.   Anderson,D.M.W. and Stoddart,J.F. (1966) *Anal. Chim. Acta*, **34**, 401.
4.   Barth,H.G. (1980) *J. Chromatogr.*, **18**, 409.
5.   Haller,E. (1968) *J. Chromatogr.*, **32**, 676.
6.   Roumeliotis,P. and Unger,K.K. (1979) *J. Chromatogr.*, **185**, 445.
7.   Roumeliotis,P. and Unger,K.K. (1981) *J. Chromatogr.*, **218**, 535.
8.   Herman,D.P. and Field,L.R. (1981) *J. Chromatogr. Sci.*, **19**, 470.
9.   Kopaciewicz,W. and Regnier,F.E. (1983) *Anal. Biochem.*, **133**, 251.
10. O'Hare,M.J. and Nice,E.C. (1979) *J. Chromatogr.*, **171**, 209.
11. Sasagawa,T., Okuyama,T. and Teller,D.C. (1982) *J. Chromatogr.*, **240**, 329.
12. Geng,K. and Regnier,F.E. (1984) *J. Chromatogr.*, **296**, 15.
13. Shelton,J.B., Shelton,J.R. and Schroeder,W.A. (1984) *J. Liq. Chromatogr.*, **7**, 1969.

14. Tempst,T., Woo,D.D.-L., Teplow,D.B., Aebersold,R., Hood,L.E. and Kent,S.B.H. (1986) *J. Chromatogr.*, **359**, 403.
15. Gurley,L.R., Prentice,D.A., Valdez,J.G. and Spall,W.D. (1983) *J. Chromatogr.*, **266**, 609.
16. Venter,J.C. (1982) *Pharmacol. Rev.*, **34**, 157.
17. Cuatrecasas,P. (1970) *J. Biol. Chem.*, **245**, 3059.
18. Lowe,C.R. and Dean,P.D.G. (1974) *Affinity Chromatography.* Wiley-Interscience, London.
19. Travis,J. (1976) *Biochem. J.*, **137**, 301.
20. Bjork,L. and Kronvali,G. (1984) *J. Immunol.*, **133**, 969.
21. Akerstrom,B. and Bjork,L. (1986) *J. Biol. Chem.*, **261**, 10240.
22. Becker,C.R., Efcavitch,V.W., Heiner,C.P. and Kaiser,N.F. (1985) *J. Chromatogr.*, **326**, 293.
23. Perrone,P.A. and Brown,P.R. (1984) *J. Chromatogr.*, **317**, 301.
24. Coplan,M. and Regnier,F.E. (1984) *J. Chromatogr.*, **286**, 339.
25. Fritz,F.E. (1987) *Biochemistry*, **17**, 1257.
26. Salas,C.E. and Selligan,O.Z. (1977) *J. Chromatogr.*, **133**, 231.
27. Wheals,B.B. and White,P.C. (1979) *J. Chromatogr.*, **176**, 421.
28. Verhaar,L.A.Th. (1984) *J. Chromatogr.*, **248**, 1.
29. Lochmuller,C.H. and Hul,W.B., Jr (1983) *J. Chromatogr.*, **264**, 215.
30. Majors,R.E. (1977) *J. Chromatogr. Sci.*, **15**, 334.
31. Martin,M. and Guiochon,G. (1972) *Chromatographia*, **10**, 144.
32. Siant-Yreix,A. and Amouroux,J. (1982) *J. Bull. Soc. Chim. Fr.*, **14**, 225.

CHAPTER 2

HPLC instrumentation

R.W.A.OLIVER

—1. INTRODUCTION

The starting point for this chapter is that of the definition of HPLC (high-performance liquid chromatography) adopted by the author; namely, that this is a method of separation of the components of a solution effected by the chromatographic process involving a column of solid stationary phase material and a liquid mobile phase in which all of the various components of the system have been designed so as to optimize the separation. In this context, optimization means first and foremost the complete (quantitative) separation of the components but it can also include such separations that are rapid and which have been automated. The key word in this treatment of instrumentation for HPLC is that of design.

This chapter is composed of three main parts; in the first part (Section 2), the basic design principles of *any* analytical HPLC system will be presented and discussed. The second part (Section 3) consists of a detailed discussion of the way in which these design principles have been converted by the instrument manufacturers into high-performance chromatographic systems designed to separate small molecules. This order of presentation has been adopted since it affords a base for the subsequent comparative discussion of HPLC systems suitable for macromolecules given in Section 4. Throughout this chapter the treatment of the technical subject matter will be aided by diagrams and photographs since it is recognized that most of the readers of this text are likely to be either biochemists or chemists rather than physicists, or mechanical or electronic engineers. However, wherever possible key references to particularly detailed studies of the design principles of the various components of modern HPLC instrumentation will be given so that the interested reader may be able to satisfy his/her curiosity.

2. HPLC INSTRUMENTATION: DESIGN PRINCIPLES

The heart of any HPLC system is that of the column of support material since this effects the initial separation of the sample mixture into its various components. Therefore, in this section, attention will be first concentrated on the way in which the development of the theoretical understanding of the basic chromatographic processes which occurred in the 1940—1960s led to the design and subsequent manufacture of the efficient column support materials now available. Further it will be shown how this advance led in turn to the design of the remaining 'off-column' parts of the modern HPLC system namely, the instrumentation. In passing it is probably correct to note that the development of

HPLC instrumentation affords an excellent example of the truth of the well known statement 'theory guides, experiment decides'!

The aim of a modern chromatographic experiment may be readily summarized in words and is as follows. Namely, it is to achieve in a reasonable time interval (minutes not hours) a quantitative separation of all of the components of a solution, which is applied to the top of the column, during its passage down the column, a process effected by elution using a suitable solvent or solvent mixture. Thus the individual components of the applied sample solution should emerge separately from the bottom of the column if the separation is complete. Now this separation process involving the elution of a solution applied to a column of permeable solid support material was first discovered by M.Tswett some 85 ago! (1). Actually, Tswett did not elute the separated components of his mixture of plant pigments off the column (of powdered chalk) since he extruded the column intact from the glass tubing in which it was contained. However, this historic experiment has been repeated recently in the elution mode (2) and it confirms that Tswett was indeed the discoverer of column chromatography. For some 30 or so years after this separation process was first published it was lost (or ignored) before it was re-discovered and used by a few workers in the 1930s in the same empirical fashion. Indeed, even until the mid 1960s column chromatography was rarely employed because it was still being operated in the classical mode. Thus, long column lengths (1 −2 m, diameter 2.5 −5.0 cm) composed of mechanically sieved particles, most commonly of silica gel or alumina, of size range 125 −250 μm, were made up in glass tubing. The sample mixtures (volume, 1 − 10 ml) were then applied manually to the top of the column and then eluted using a suitable solvent under gravity. Further, the progress of the separation was monitored by manually collecting fractions of the column eluate (volume, 1 − 10 ml), which were then sequentially analysed by some discrete method. Under these conditions the overall time taken to complete the quantitative separation of the sample mixture was measured in hours if not days and required considerable amounts of sample, solvents and operator time.

Since the resurgence of interest in column chromatography in the late 1960s the experimental situation has changed completely. The chromatographer now has available a wide range of HPLC analytical systems which are completely automated and which can continuously quantitate the separation of μl volumes of a wide range of sample mixtures achieved within a time scale of some 2 −30 min. This considerable technical achievement has resulted primarily from the success of those research and development studies aimed at understanding the mechanisms of the separation processes operative during the chromatographic elution. These theoretical studies may be considered to originate in the work of Martin and Synge (3) and they have been well summarized by Synder and Kirkland (4). Most of the predictions of these theoretical studies were tested by performing quantitative separations on mixtures of small molecules. However, of particular interest to readers of the present text is that during the last decade, the theoreticians have also been studying the separation of macromolecules [see recent review by Synder and Stadalius (5)]. The most important result of the theoretical studies has been that quantitative measures of the column variables, namely particle size, particle pore size (if porous), column length, column diameter for the different types of column chromatographic supports, for example normal phase, reversed-phase, ion-exchange, hydrophobic interactions, size-exclusion materials, have been listed. Further, the relationships between these variables and the quantitative efficiency of the separations achieved under a range of operating conditions, viz. volume of sample applied, solvent flow-rate, pressure drop, mode of elution-gradient/isocratic and detector, have been elucidated. As a direct result of these studies the research chromatographers have been able to list specifications for high-performance columns and for those optimum operating conditions under which efficient separations may be achieved (6).

Reference has already been made to three key research and development chromatographic studies (4−6) and therefore only a summary of the major conclusion of these studies, followed by a discussion of the practical consequences which arise from it, will be given here.

The major conclusion of these studies is that HPLC requires first and foremost that the chromatographic column should be composed of uniformly packed, microparticulate support materials ($d_p \le 10\ \mu m$) of uniform particle size.

The practical realization of this conclusion required considerable technical effort and expertise. Unfortunately, because of the enforcement of secrecy for commercial reasons, this technological achievement has not been well documented in the open literature and few details are available. It is for this reason that most chromatographers, with the possible exception of those using high-performance affinity chromatographic columns, either purchase column support materials from the manufacturers and pack them into columns of the required dimensions or they purchase ready made columns. It should perhaps be noted that some column support materials, especially the rigid size-exclusion chromatographic supports cannot be purchased separately, possibly because the manufacturers wish to recoup the research and development costs even though it may be that these particular materials are difficult to pack correctly.

The remarkable degree of success achieved by the manufacturers of HPLC column support materials may be gauged by a considered reading of the following precise and practically quantitative description of one such material, chosen at random by the present author from the Macherey-Nagel catalogue. Thus:

> 'Nucleosil® is a spherically shaped totally porous silica. We manufacture this packing material with different pore diameters (50, 100, 120, 300, 500, 1000 and 4000 Å) and pore volumes between 0.65 and 1.0 ml/g. Particle sizes for analytical purposes range from 3 μm to 10 μm with very narrow fractionation. All of the narrow pore Nucleosil packings (50−120 Å) are stable up to 600 bar (8500 p.s.i.). The wide pore packing are stable up to 400 bar for 300 and 400 Å pores, to 300 bar for 1000 Å and 400 Å material.'

Further, when it is realized that the surface of these carefully manufactured support materials may be subsequently subjected to controlled chemical modification, a process which results in the formation of a wide range of completely new surfaces with differing chromatographic properties, it is possible to understand why the chromatographer is nowadays presented with a bewildering choice of stationary phases. For assistance with this particular problem the reader is referred back to Chapter 1.

Having established the fact that a wide range of HPLC column support materials and packed columns are currently available, the conditions under which the latter must be operated will now be discussed in general terms, so that the reader may be able to understand how the modern, modular HPLC system came to be developed.

'The main operational requirement for an HPLC column is that it should be eluted under pressure.'

This dictate stems from a combination of the fact that HPLC columns must be composed of small particles in order to achieve high resolution separations and from the insistence (of the chromatographer) that the separations be completed within a short time period (2−30 min). Thus in practice that eluant must be forced through the column of small particles at reasonable flow-rates. All HPLC systems must therefore possess an elution module which can be operated under pressure and because of this, HPLC was at first designated as high-pressure liquid chromatography. As a consequence of this operational practice all HPLC support materials must be able to withstand the range of pressures necessary to generate the optimal flow-rates (e.g. see previous description of pressure stability of Nucleosil®). It should however be recognized that the actual pressures needed depend primarily upon the eluant flow-rates chosen in order to achieve the separation in a given time.

The actual pressures required depend upon the following variables, namely particle size, column dimensions and, to some extent, upon the nature of the particular eluant(s). This key operational requirement of elution under pressure also accounts for the practice of packing HPLC columns in metal rather than glass tubing and of de-gassing the eluant before pressurizing it in order to prevent the formation of air bubbles with subsequent physical damage to the columns. Further, the need to operate HPLC columns under pressure obviously complicates the process of applying the sample and, because the columns are enclosed in metal, of monitoring the progress of the separation. A considerable amount of research and development effort has been employed on these two key areas of HPLC instrumentation and as a result a variety of well designed and engineered sample application and analyte detector modules are now readily available. Finally, to complete this introduction to the basic principles of HPLC instrumentation one final module must be considered. This consists of the electronic control module, which is designed so as to control all of the previously discussed physical function modules, perform all of the necessary calculations and to drive the output device. It should perhaps be noted that this module is only really essential if the separations performed with the HPLC system need to be automated. A number of scientific instrument manufacturers now offer such completely automated modular analytical HPLC systems (see Appendix for a list of names and addresses) and *Figure 1* is a photograph of one such system. It is hoped that the reader who studies this figure will, with the aid of the explanatory key, be able to identify each of the individual modules discussed in this section.

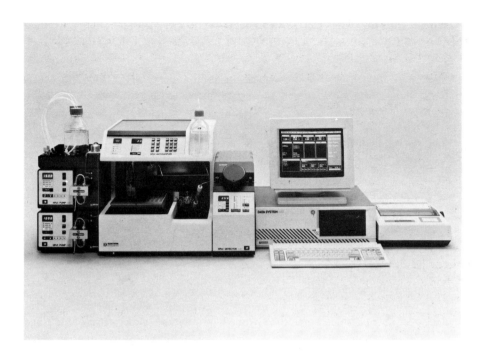

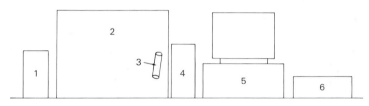

Figure 1. Photograph and block diagram of a modern, automated modular HPLC analytical system (Kontron Instruments-HPLC system 400). (**1**) The eluant delivery module (dual piston pumps); (**2**) sample application module (automated); (**3**) HPLC column module; (**4**) analyte detector module (UV-VIS dual wavelength detector); (**5**) central control module (controls module 1−4 and 6); (**6**) data output module (printer-plotter).

Automated HPLC systems, such as that shown in *Figure 1*, are capable of working 24 h/day, 7 days/week for 52 weeks of the year, and because of this they have already completely transformed many areas of routine analytical biochemistry. In the following section the key specifications and design principles of each of the component modules of a modern automated HPLC system suitable for the analysis of small molecules will be presented and discussed in the hope that this will give the reader the information necessary to judge the individual modules of any particular system.

3. MODULAR HPLC INSTRUMENTATION FOR SEPARATION OF SMALL MOLECULES: DESIGN PRINCIPLES AND SPECIFICATIONS FOR EACH MODULE

In this section, the various modules which make up a modern HPLC system will be discussed in the order in which they have been introduced previously and labelled in the legend to *Figure 1*.

3.1 **The eluant delivery module**

The separation of a mixture of small molecules applied to a HPLC column may be achieved by eluting the column with either an eluant of constant composition (isocratic mode) or one with changing composition (gradient mode) at a constant flow-rate generated by pressurizing the eluant. Thus this module in general consists of three distinct components: firstly, the solvent(s) reservoir(s); secondly, the gradient forming system; and thirdly, the high pressure pump(s). The essential design requirements for the latter component will be considered first since the pump forms the key part of this module. It should be remembered from the previous section that the primary function of the pump is to enable the chromatographer to force the eluant through the HPLC column at a constant flow-rate in order to achieve the desired separation in a short time interval. Thus the operating parameter governed by the high-pressure pump is that of the eluant flow-rate so the pump designer needs to know first and foremost the range of eluant flow-rates required. The top of this range is determined by the maximum pressure which can be generated by the pump and withstood by the HPLC column. As a result of much research and development work the range of flow-rates encountered with various HPLC columns in general analytical use is now known and is 10 μl/min for microbore columns (i.d. ≤ 2 mm) and up to 10 ml/min for analytical columns (i.d. $\simeq 4-5$ mm) with the most widely used flow-rates for the latter being between 1.0 and 2.0 ml/min. To obtain such flow-rates with modern HPLC analytical columns (of up to 25 cm in length) the eluant may need to be pressurized up to 600 bar (8500 p.s.i.).

In practice, the desired flow-rate is chosen, after first limiting the maximum pressure drop to that which the column can withstand, and often the corresponding pressure generated is not even recorded. However, it is these two sets of figures, that of flow-rate and maximum pressure, which together form the key design specifications for the pumps used in HPLC.

One further set of figures are also of interest to the pump designers. These are the long term and short term fluctuations in the pressure in these solvent delivery modules. The first of these tend to originate in physical changes in the HPLC column itself—thus columns 'block up' with continued elution so that if the pressure generated by the pump remains constant the flow-rate decreases, or alternatively, if the flow-rate is maintained constant then the generated pressure must be increased. Clearly it is difficult to quantify these fluctuations since some columns are good, whilst others are bad and it is thus sound practice to monitor the operating pressure at any chosen flow-rate as a function of time in order to classify any given column. The short term fluctuations tend to originate in the method of generating the pressure, that is in the design of the pump itself and the design principles of the latter will now be presented and discussed.

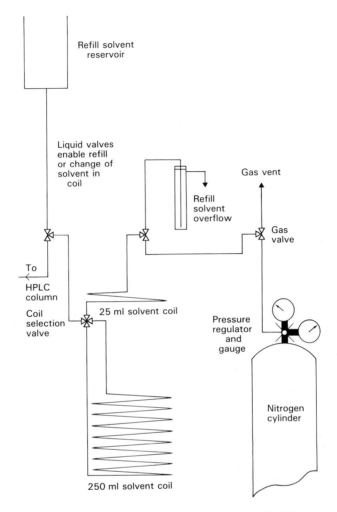

Figure 2. Schematic diagram of pulse-free, coil-type, pneumatic pressure (0−2500 p.s.i.) pump system.

Liquid pressurizing systems can be classified into four types, namely (i) those which have no 'true' pump—the pneumatic pumps; (ii) those with reciprocating piston pumps, (iii) those with syringe type pumps, and finally (iv) those with a combination of the first and second systems, namely pneumatic amplifier pumps. Of these various types the first two (cheaper ones) will be discussed here because they are the more popular and because the syringe type pumps have been discontinued by most HPLC pump manufacturers in favour of the modern reciprocating pump.

3.1.1 *Pneumatic 'pumps'*

These simple, inexpensive systems employ pressurized gases to drive the eluant down the HPLC column. The maximum pressure achieved by such systems are thus limited by the available gas cylinder pressure (normally 175 bar, 2500 p.s.i.). The gas, usually

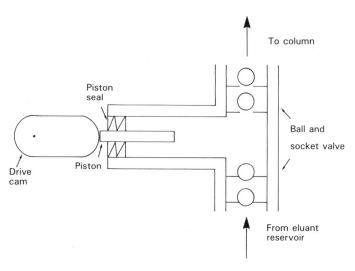

Figure 3. Schematic diagram of a single piston, mechanical, constant volume, reciprocating HPLC pump.

nitrogen, is generally applied via a pressure regulator either directly to the eluant surface or, in some configurations, to the eluant via a diaphragm of suitable material. Now, although these simple systems can only pressure a limited volume of solvent, they will deliver a completely pulse-free flow of eluant at constant pressure. The actual constancy of the volumetric flow is thus dependent upon the column maintaining constant flow resistance during the separation period and this is normally achieved in practice.

Figure 2 is a schematic diagram of the author's home-made system in which the gas pressure is applied directly to the surface of the eluant which is maintained in a stainless steel coil. A number of the latter, of differing lengths are available giving a range of solvent volumes, but they are all of relatively small internal diameter so that only a small surface area of the solvent is open to the pressurizing gas (N_2), in order to minimize the effect of the gas dissolving at the high-pressure end and bubbling out at the low-pressure end of the system (bottom of the HPLC column) and hence into the detector cell!

In spite of the simplicity of this type of eluant pressurizing system the present author has employed it with considerable success over a period of 15 years not only for teaching but also for non-routine analytical research work.

It should perhaps be noted here that these pneumatic 'pumps' can however only operate for a limited time before re-filling of the solvent reservoir is necessary. Further, whilst they are excellent devices for operating HPLC columns in the isocratic mode they are not suitable for gradient mode elution. In spite of these drawbacks, the simplicity and cheapness of this type of eluant-pressurizing system make them very attractive to a number of laboratories and especially those on limited budgets.

3.1.2 *Reciprocating piston pumps*

These pumps utilize small volume (ml) hydraulic chambers fitted with reciprocating pistons to generate the pressure required to drive the eluant through the HPLC column

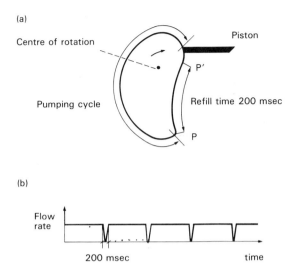

(a)

Centre of rotation

Piston

Pumping cycle

P′

Refill time 200 msec

P

(b)

Flow rate

200 msec

time

Figure 4. (a) Design details of the drive cam and (b) resultant liquid flow-rate profile in a modern single reciprocating piston pump.

which is attached in line. Two ball and socket check valves are synchronized with the piston drive to allow alternate filling and emptying of the eluant from the hydraulic chamber and the mechanism of action of the pump is indicated in *Figure 3*.

As can be seen from a consideration of *Figure 3* during the withdrawal stroke of the piston the pressure must fall considerably and as a result of this the pump generates a series of pressure pulses.

Hence during the last decade the pump manufacturers have paid particular attention to the design of the shape of the mechanical cam which drives the piston so that the delivery stroke is of greater duration than the refill stroke. As a result of this re-design the piston operates throughout its entire length of positive travel (insertion) and the liquid flow-rate is varied by controlling the frequency of the reciprocating piston. In one such pump (Kontron 420) the shape of the drive cam, shown in *Figure 4a*, leads to a refill time of about 1/6th of the positive pumping time and hence to the pulsed eluant flow-rate profile shown in *Figure 4b*.

Even with such careful design of the cam drive mechanism, single reciprocating piston pumps produce significantly pulsed eluant flow-rates and hence some dampening or other appropriate course of action must be taken by the instrument maker to overcome this limitation since otherwise they would not be suitable for use in HPLC.

In a simple (low cost) HPLC system using a single reciprocating piston pump it is common practice to install a pulse damper in order to achieve a smooth eluant flow-rate. To do this a combined expansion restriction device is employed — a coiled tube (Bourdon tube) or pressure gauge provides the expansion volume and this is followed by a capillary restrictor. Unfortunately this solution to the experimental problem requires that a large resistance be used so that a considerable pressure build up occurs within the pulse dampener. In turn this leads to a marked lowering of the useable pressure

51

Table 1. List of modern HPLC dual piston reciprocating pump operating specifications.

Range of flow-rates	10 μl/min to 20 ml/min
Accuracy of flow-rates	$\leq 1\%$
Precision of flow-rates	$\leq 0.3\%$
Drift	$\leq 1\%/10$ h
Working pressures	Up to 600 bar (8000 p.s.i.)

generated by the pump for HPLC, a fact which restricts the use of the pump at high flow-rates. In an attempt to overcome this limitation some pump manufacturers use a high-pressure metering valve as a variable resistance in place of the capillary restrictor unit of the pulse dampener. This may then be set to give either minimum pulsation or minimum pressure drop according to the particular HPLC requirements.

However, the eluant flow-rate pulsations are most readily reduced without loss of pumping ability by the use of two (or more) reciprocating pistons which are linked in parallel but out of phase. Thus in the latest survey of HPLC equipment to be published by McNair (7) all of the 31 HPLC systems studied were reported to use reciprocating piston pumps, only one of which employed a single piston pump (Varian). One system used a triple piston pump (Anspec), the remaining HPLC systems employed double piston pumps and only three of these required pulse dampeners in line. The principle of the dual piston pumps is that the two pistons are mounted 180° out-of-phase on the same drive cam so that one piston is withdrawing whilst the other piston is delivering eluant to the chromatographic column. This basically simple arrangement reduces the pressure fluctuations to less than 5% of the total back-pressure and again by careful design of the cam drive mechanism excellent specifications, such as those given below for the Kontron 420 HPLC dual piston reciprocating pump, are now quoted by a number of manufacturers. A list of operating specifications is given in *Table 1*.

Any such dual piston pump with the specifications listed in *Table 1* is most suitable for HPLC since the pulsations have been reduced to a level comparable to the signal/noise ratio of even the most sensitive of HPLC detectors. Further such pumps may be used as part of a gradient-forming elution system. It would thus seem that the pumping problems of the HPLC chromatographer have been solved by the design and production efforts of the instrument makers. It should however be recognized that in practice, the limiting feature of these modern pumps is the requirement for the inlet and outlet ball and socket valves (*Figure 3*) to be made and to be kept perfect! So even if the valves have been made correctly, it is still possible for a speck of particulate matter or an air bubble to get trapped inside them, thus severely limiting the performance of the pump. For this reason alone one should only use solvents which have been de-gassed and filtered through a 0.2 μm filter before use. Even so it is sometimes found necessary in practice to have to strip down and clean the various parts of the pump. Clearly if this can be done 'in-house' it is likely to be faster and certainly cheaper than having to call in the manufacturer. The ease of maintenance of the pump is thus a factor of key importance in the choice of pump. Some idea of the magnitude of the task involved may be gained by the reader on examination of *Figure 5*, which shows all of the removeable and hence cleanable parts of a modern dual reciprocating piston pump (Kontron 420).

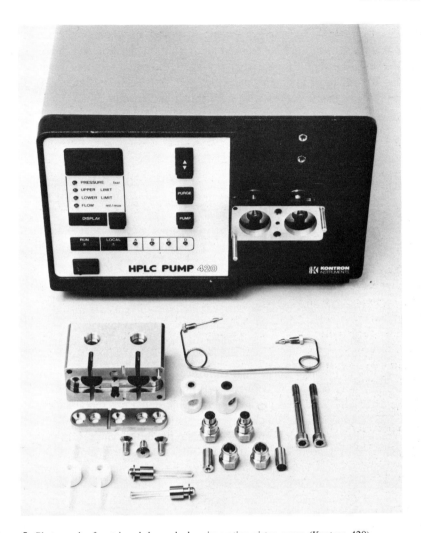

Figure 5. Photograph of a stripped down dual reciprocating piston pump (Kontron 420).

It should perhaps be noted that many pump manufacturers are aware of the importance of easy access and re-assembly of this vital piece of HPLC equipment and have taken the necessary design action! Finally on the subject to reciprocating pumps, mention should also be made concerning the materials used to construct the pump. The pistons may be constructed of borosilicate glass, sapphire, chromium plated stainless steel or ceramic, the gaskets and seals are usually made of Teflon or Kel F and the metal parts are generally made of 316 stainless steel. The choice of materials employed is dictated not only by the resistance to wear but most importantly by the solubility properties of the solvent or solvent mixtures used as eluants, that is by the nature of the HPLC separations. Thus the choice of an HPLC pump is also determined by the latter. As will be discussed later in Section 4, the separation of certain large biologically active molecules, for example enzymes, require that the pump materials should be chemically inert so that trace metals are not leached out from those surfaces which are in contact with the eluant. Further, in ion-exchange HPLC, the pump materials must be able to withstand the range of ionic solvents employed (see Table 3, Chapter 4). Because of these problems many reciprocating piston pump manufacturers now include a piston flushing facility as standard in order to prolong the life of the piston seals and this detailed design feature is to be commended.

3.1.3 *Gradient-forming systems*

Elution may be performed in HPLC in either the isocratic mode (eluant of constant composition) or in the gradient mode (eluant of changing composition). Many HPLC separations involving simple mixtures of small molecules may be achieved in the isocratic mode and this elution mode is always the first choice since all that is required is one reciprocating dual piston pump or a single pneumatic 'pump'. However, for difficult HPLC separations gradient-elution is necessary and hence a gradient-forming system is required. These are of two types: (i) the externally generated gradient system in which the solvents are mixed on the low-pressure side of the pump at atmospheric pressure and (ii) the internally generated system in which the solvents are mixed on the high-pressure side of the pump. In gradient-elution the mixing system must respond rapidly to solvent changes and hence the volume of the solvent-mixing chamber and the connecting volume from this chamber to the injection part of the HPLC column must be kept as low as possible. Since dual reciprocating piston pumps have low chamber volumes they may be used in both types of gradient-forming systems.

In the externally generated gradient system, the pump external eluant reservoir is the mixing chamber for the mixture of solvents (see *Figure 6a*). The advantages of such a gradient system are low cost and complete flexibility with regard to the number of solvents and to the composition of the mixtures. However, the major disadvantage of this type of gradient system is that it is very time consuming and inconvenient since the mixing chamber must be cleaned with fresh solvent for every gradient.

By contrast, the internally generated gradient system which employs two (or more) HPLC dual piston reciprocating pumps to force the solvents at high pressure into a mixing chamber prior to the HPLC column; *Figure 6b*, can quickly generate eluant mixtures of varying composition, has rapid turn around times and above all is amenable to automation. In such systems the mixing chamber has a small volume (0.5–2.0 ml), is made of chemically inert material and is designed so as to ensure adequate mixing of the solvent mixture and to be cleanly swept when required.

Further, the HPLC instrument manufacturers have gone to considerable trouble to automate these internally generated gradient-forming systems by involving computers and programmers, but unfortunately this has led to such systems becoming prohibitively expensive for some laboratories. In spite of this fact these automated systems are recommended, especially for use in research and development laboratories. It is also recommended that the internally generated gradient-forming system should be regularly checked for accuracy by the technique of placing the same solvent in both pump reservoirs, *Figure 6b*, and then adding a small amount of a good UV absorbing compound, for example benzoic acid to *one* of the solvent reservoirs. The gradient

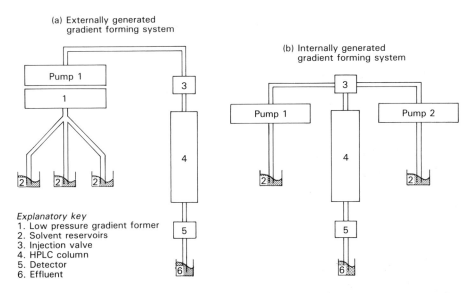

Figure 6. Schematic diagram of the two types of gradient-forming systems. (**a**) Externally generated gradient-forming system; (**b**) internally generated gradient-forming system.

profile should then be recorded using a column filled with glass beads connected to a 254 nm detector. By this means the 'electronic profile' may be compared to the actual off-column profile at the detector and in particular one may observe the beginning and end of a gradient run where one pump is phasing in or out at very low flow-rates.

3.1.4 *The solvent(s) reservoir(s)*

Ideally these reservoirs should meet the following specifications.

(i)   They must hold a volume sufficient for repetition analysis.

(ii) They must provide for solvent de-gassing either by heating or applying vacuum or by sparging with helium.

(iii) They must be chemically inert with respect to the particular solvent.

These specifications are realized in the present author's laboratory by using glass containers of 500 ml to 2000 ml volume which are carefully capped once the solvent has been de-gassed (under partial vacuum). Finally, before the HPLC grade solvents are transferred to the pumps they are all filtered through a 0.2 μm stainless steel filter.

3.2 **The sample application module**

This module is the device employed to load the sample solution onto the column. In all forms of chromatography, this loading process is critical for efficient separation and the HPLC technique is no exception to this generalization. The specification for an ideal sample application module is as follows. Namely, it is a device which enables the user to apply in a reproducible and convenient manner a range of sample volumes (of the order of microlitres for analytical separations) onto the pressurized HPLC column as a discrete plug without loss of efficiency of the column. Further, if possible, the device should be capable of being automated.

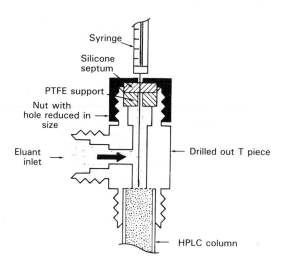

Figure 7. Diagram of microsyringe – septum injector.

No such 'ideal' device has yet been designed but some are believed to approximate quite closely to it. Indeed it is difficult in practice to measure experimentally the degree of deviation of any device from the ideal as defined above. Thus, if such an ideal sample application module existed, then a perfect loading of an aliquot of a sample solution into a very short tube connected to a sensitive detector on the outlet side and a pulseless HPLC pump on the inlet side would lead to a rectangular 'peak' being produced on the strip chart recorder. In practice, only trailing peaks are obtained—a finding which may be explained in part by the rapid production of a needle-like flow pattern in the injected slug of liquid when it is placed in the eluant flowing at normal HPLC flow-rates (8). Therefore in the account of the various devices which follows, emphasis will be placed on the practicalities of each type. Basically two types of sample application modules have been designed for use in HPLC. These are (i) microsyringe – septum injector devices and (ii) septum-less micro sampling values.

3.2.1 Microseptum injector

These devices are the cheapest and simplest of the sample injection devices and may even be home-made but since they suffer from a number of serious disadvantages they are not widely used today. *Figure 7* is a diagram of one of the simplest designs of these devices and also indicates its mode of action. With care, this device is capable of giving good results, especially for injections made onto the column, since the dead volume is zero (as shown in *Figure 7*) and in particular if the eluant flow is momentarily stopped during the injection procedure (stopped-flow injection). The major problems which arise are that the septa, usually made of silicone rubber tend to leak after about 20 – 30 injections and they may also contaminate the column by physical breakage with subsequent dissolution of the pieces in the HPLC eluants. In addition they may only be used at pressures up to about 10 – 15 MPa (1450 – 2175 p.s.i.). Further, because

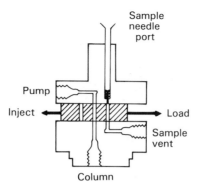

Figure 8. Schematic diagram of septum-less syringe loading microsample valve (Rheodyne model 7520).

these devices cannot be automated they are tedious to use in routine analysis and hence quite considerable errors of reproducibility may arise.

3.2.2 *Sampling valves*

By contrast with the previous device the modern HPLC high pressure valve cannot be home-made since they are complex pieces of equipment. They operate quite simply by transferring an aliquot of the sample solution at atmospheric pressure from a syringe to a sample loop. This loop is then connected by switching of the valve to the high pressure eluant stream, which carries the sample onto the column. The sample loop may be loaded by a complete filling or a partial filling method and the pros and cons of each method has been well documented (9). The almost uniform adoption of these devices by HPLC chromatographers is due not only to the design excellence but also to the careful choice of the materials of construction and to the sophistication of modern precision engineering practice. Because of the complexity of the manufacture it is not therefore surprising to find that a small number of firms, for example Rheodyne, Valco, Hamilton exist for which such devices form the major product. Rather than give a detailed account of each of the individual HPLC sample valves now available—the above firms produce excellent technical and descriptive notes which are supplied free upon request, for addresses see Appendix—only one such valve, that manufactured by Rheodyne, will be described in order to illustrate the technical manufacturing problems which have been overcome. This model has been chosen because it is recommended for use with microbore columns (i.e. ≤ 2 mm), uses sample volumes of 0.2, 0.5 and 1 μl only and is therefore at the forefront of valve technology.

Figure 8 is a schematic diagram of this Rheodyne model 7520 syringe loading sample injector.

The valve uses a small hole, of volume 0.2, 0.5 or 1 μl, drilled in a flat rotor (hatched section, *Figure 8*) which serves as the internal sample 'loop'. Further, this rotor is sandwiched between the two stators as shown. This sample 'loop' is loaded, as shown in *Figure 8*, by inserting the correctly shaped syringe needle into the built-in needle port and injecting the appropriate aliquot of sample solution. The space between the

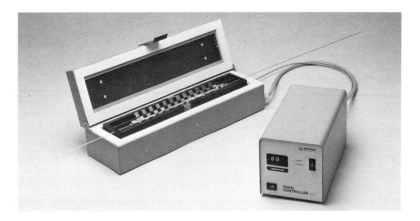

Figure 9. Typical HPLC column oven (Kontron).

square needle tip and the rotor surface contains only 0.3 μl. Rotating the valve handle (not shown in *Figure 8*) places the sample volume contained in the 'loop' into the eluant stream and hence onto the HPLC microbore column.

The adoption of these valves stem from the fact they are septumless and hence suffer from none of the previously discussed disadvantages of septa, that they operate at pressures of up to 30−40 MPa (4350−5800 p.s.i.) and may be automated. The author is convinced that the problems of manual sample application in analytical HPLC have thus been solved by the valve manufacturers.

Further, as the popularity of the HPLC technique for routine analytical work has increased, the instrument makers have invested much research and development effort into automating the technique. As a result most of the larger HPLC manufacturers offer particularly versatile automatic sample application modules which are generally built around a pneumatically actuated or motorized sample valve. One such system is shown in *Figure 1*, module 2. These tend to operate as follows—on command from the control system (*Figure 1*, module 5) an appropriate aliquot of the sample solution is transferred to the valve from a pre-determined sample contained in capped sample vials. This process may be repeated as often as may be required at certain specified, but variable, time intervals. Whilst much of the undoubted success of these automatic sampling systems depend on the electronics, good quality design and care in the production of the mechanical side of the system is also essential. Unfortunately, both of these properties of the device cost money but for most routine quality control laboratories such automated sample application devices are probably cost-effective.

3.3 **The HPLC column module**

As stated in the introduction to this chapter the HPLC column forms the heart of any HPLC system. In practice, its careful design and the correct choice of the type and quality of its packing material are essential for good separation and minimal dispersion of the sample components and this forms the subject of Chapter 1. In general, for the reasons explained earlier, stainless steel columns of internal diameter 4.0−4.6 mm and length 10−30 cm, with minimal dead volume end fittings, are employed to contain the column packing materials used in analytical systems. It is a common and sound practice to insert a 'guard column' between the sample application module and the analytical columns in order to protect the latter from contaminants originating from either the eluant or the sample. Guard columns are designed to be easily replaced or re-packed with a minimum expenditure of time and expense. Normally a short (5 cm) stainless steel column of the same internal diameter as the main column is employed which contains a pellicular support with a similar chemical functionality to the main microparticulate column (see Chapter 1). Care must be taken to use small diameter stainless steel tubing (i.d. ≤ 0.25 mm) to connect both ends of the guard column, one to the sample application module and the other to the analytical column. Further the ends of all of the tubing employed must be cut square in order to obviate extraneous 'dead' volume and the lengths must be kept to a minimum.

In terms of instrumentation, the only other part of the HPLC column system that requires consideration here is the temperature control module of the column. Most of the early workers obtained high-performance separations at room temperatures with the columns exposed to the laboratory as shown in *Figure 1*. However, by 1969, Maggs (10) had concluded from his studies that it was necessary to control the column temperature to within ±0.2°C if *repeatability* of retention volumes was to be maintained within 1%. The fact that many workers have not thermostatted their HPLC columns and yet have obtained reproducible separations is probably due to the fact that many air-conditioned laboratories maintain ambient temperature to within ±1°C. Since the majority of HPLC separations are performed at room temperature, column temperature control modules are therefore still regarded by many HPLC manufacturers as an optional extra (7) and one such is shown in *Figure 9*.

The design specifications of these column ovens are that they maintain the column temperature constant at +0.2°C, at a pre-set temperature, within the range 20 – 150°C. The ability to perform HPLC separations at high temperatures (both bonded reversed-phase and ion-exchange columns may be operated up to 65°C) is often advantageous since the sample solubility increases, the rate of mass transfer increases and hence the column efficiency increases. Accordingly for applications which require chromatographic reproducibility and that the columns be operated at maximum efficiency, HPLC column ovens should be considered essential.

3.4 Analyte detector module

In this discussion only continuous HPLC monitoring or on-line devices will be considered since the discrete methods referred to earlier in the introduction involving sequential analysis of collected eluate fractions are laborious, time consuming and are not therefore used today. An on-line HPLC analyte detector module is defined as a carefully designed device, through which the eluate from the HPLC column flows, and which (in most cases) generates a continuous electrical output signal that is a function of the mass of the analyte or of the concentration of the analyte in the mobile phase.

In most HPLC systems, this electrical signal is passed directly to a strip chart recorder to provide a permanent paper record of the progress of the separation as a function of time, or the HPLC chromatogram. Now the careful choice and design of this module is essential because even though the separation process is achieved on the HPLC column, the separation is monitored 'off-column' and assessed from the resultant chromatogram. Therefore, in addition to the column, the quality of the separation achieved, as measured from the chromatogram, depends upon the detector response time, the analyte electrical response signal to detector noise ratio and the flow-cell design. The latter 'flow-cell' is that part of the off-column conduit system adjacent to the bottom of the column which contains the portion of the eluate which is actually being monitored by the detector device.

Because of the importance of this module a really vast amount of research and development time and energy has been spent on trying to develop an 'ideal' universal detector – flow-through cell combination. The specifications for such an analyte detector module are as follows.

(i)   The detector should give a rapid (msec) electrical response signal to all analytes.

(ii)   At a detector analyte response signal to electronic noise ratio of 2, the sensitivity should be similar to that used in gas chromatographic detectors, that is, of the order of $10^{-11} - 10^{-12}$ g/ml.

(iii)   The detector analyte response signal should be linear over an analyte concentration range from the minimum (corresponding to the above sensitivity value) to a maximum 10^5 to 10^6 times greater than this minimum concentration.

(iv)   The detector electrical response signal should be independent of the nature and composition of the eluant, its temperature, flow-rate and pressure.

(v)   The flow-through cell should be as close as possible to the bottom of the analytical column in order to minimize analyte band dispersion effects caused by the parabolic flow of the eluate once it is 'off-column' (8).

(vi)   The flow-through cell should possess carefully designed inlet and outlet tubes designed to change the geometry of the column connecting tube so as to introduce radial flow of the eluant and thus reduce the analyte's parabolic velocity profile (8).

(vii)   The flow-through cell should possess as small a volume as possible so that the volume of the eluate being monitored approximates to a very small cross-section of the flowing eluate.

In spite of the large expenditure of research and development time and energy on this topic by industry and by academia, the construction of an ideal universal analyte detector and flow-through cell module which fulfills all of these specifications still eludes us. Indeed the current instrumentation situation is that a very large number of different analyte detector modules are now available so that the chromatographer is presented with an additional problem—namely which one (or more) to choose. The names of these different detector modules are listed in *Table 2*, together with some comments concerning their applicability to particular problems.

As can be seen from a study of the three sections forming *Table 2*, the most popular of the common detectors are those involving UV photometers and because of this the remainder of this discussion of the detector modules will be confined to these. Those readers who having studied *Table 2* wish to pursue this subject in more depth, or to consider other detectors are recommended to read Scott's excellent book which is devoted exclusively to this subject (11).

Table 2. A listing and commentary on some analyte detector modules which are commercially available.

Name	Comment
Section 1. Common detectors	
1. UV photometric	
(i) (Fixed wavelength)	See text.
(ii) (Variable wavelength)	
(iii) (Diode array)	
2. Fluorescence	*High sensitivity*, but of limited applicability because so few analytes naturally fluoresce. Range of applications may be considerably increased if analytes are chemically modified to form fluorophores by either pre- or post-column derivatization.
3. Refractive index	*Moderate sensitivity*, applicable to a wide range of analytes but is sensitive to small temperature ($\pm 0.01\,^\circ C$) and pressure variations and hence can only be used for isocratic separations.
4. Electrochemical	
(i) Amperometric	*High sensitivity*, but only applicable to analytes which can be oxidized or reduced at a suitable working electrode.
(ii) Conductimetric	*Medium sensitivity*, applicable only to anions and cations with pK_a or pK_b <7. The detector most widely used for ion-exchange HPLC.
Section 2. Less common detectors	
5. (i) Fluorescence (β-induced)	*High sensitivity*, has limited range of applicability e.g. aromatic hydrocarbons.
(ii) Refractive index thermal lens (laser induced)	*Moderate sensitivity*, applicable to wide range of analytes and suitable for isocratic separations, needs further development work to improve sensitivity.
(iii) Interferometry	*High sensitivity*, applicable to wide range of analytes only suitable for isocratic separations, needs further research and development studies.
(iv) Light scattering	Detectors based on these effects are suitable for certain classes of analytes, e.g. polymers and large molecules or optically active molecules. They all need further research and development studies.
(v) Circular dichroism	
(vi) Optical rotating dispersion	
6. Radioactivity	*High sensitivity*, clearly applicable only to radioactive analytes. Expensive.
Section 3. HPLC detector – spectrometer combinations	
7. Infra-red	*Limited sensitivity*, and restricted to non-aqueous eluant systems.
8. Nuclear magnetic resonance	*Limited sensitivity*, applicable to a wide range of analytes, costly, still under development.
9. Electron spin resonance	*High sensitivity*, but only applicable to those analytes which can be induced to form free radical derivatives by pre- or post-column derivatization.
10. Mass spectrometry	*Good/high sensitivity*, applicable to analytes of low and medium molecular mass ($10^2 - 10^4$). Currently under very active development to extend mass range and to improve sensitivity. Electrospray mass spectrometric techniques promising (27).

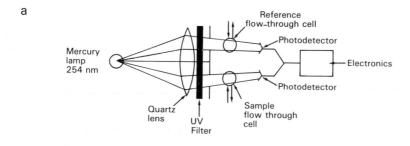

Figure 10. Schematic diagrams for **(a)** simple fixed wavelength UV photometer – flow-through cell module; and **(b)** variable wavelength double beam UV photometer – flow-through cell module.

3.4.1 *UV photometric detectors (design principles)*

These detectors may be divided into the three types shown in Section 1 of *Table 2*, namely (i) simple fixed wavelength, (ii) dispersion type (prism or grating) variable wavelength double beam UV spectrophotometers and (iii) diode array UV photometers. *Figure 10* consists of schematic diagrams of the first two types of detectors.

(i) *Simple fixed wavelength detectors*. To the left of *Figure 10a* is the low pressure mercury vapour lamp that emits its intense line at 253.7 nm together with a number of weaker spectral emission lines which are filtered out by an appropriate filter as shown. A quartz lens focuses the UV light onto the pair of flow-through cells and finally onto the two photodetectors as shown. The electrical outputs of these two detectors are passed through a pre-amplifier and thence to a log comparator to yield the output in absorbance units $[A = \log I_0/I_t]$. Because many analytes do not absorb at 254 nm, a variety of this type of detector which employ a medium pressure mercury lamp in conjunction with suitable filters is now available. Some operate at two wavelengths normally, 254 and 280 nm, others may operate additionally at 312, 365, 436 and 546 nm.

(ii) *Dispersion type variable wavelength detectors.* Some of this type of detector employed in modern HPLC systems are true recording instruments which allow the measurement of a complete UV/visible absorption spectrum on any analyte peak which is 'trapped' (by stopping the pump) in the flow-through cell. Others of this type have manual selection of wavelength within the range 260−850 nm. This manual process allows the chromatographer to select either the wavelength at which the analyte has maximum absorption or the wavelength at which any interfering substances have minimum absorption.

A study of *Figure 10b* shows the optical arrangement of a typical grating instrument— light from either a deuterium discharge lamp (UV) or a quartz halogen lamp (VIS) is focused on the entrance slit of the grating monochromator. In this design of the optics, the dispersed light is incident onto the exit mirror. By varying the mechanical position of the grating housing any desired wavelength may be focused onto the exit slit from whence it then proceeds via the beam splitter to the pair of flow-through cells and hence alternately to the detector by the operation of the rotating chopper. As before the ratios of the detector signals are converted into absorbancies using a logarithmic converter.

(iii) *Diode array variable wavelength detectors.* These detectors are of the most recent of technical advances in commercial HPLC detectors (12). As can be seen from a study of the schematic diagram given in *Figure 11*, in this system a polychromator (holographic grating) disperses the spectrum from the light source, after it has passed through the flow-through cell, across a miniaturized diode array. Consequently, the whole of the absorption spectrum of the UV absorbing analyte is monitored simultaneously on a millisecond time scale. The major advantage of this type of detector over the two discussed previously is that the UV/VIS spectrum of each chromatographic peak may be measured without disturbing the eluate flow. The major disadvantages are that they are of reduced sensitivity, because of an increase in short term electrical and optical 'noise', have a smaller linear working range and finally are more expensive than the traditional UV detector.

3.4.2 *UV photometric detectors (critique)*

These detectors, when used in conjunction with a well designed, low volume, flow-through cell, continuously monitor the absorbance of any UV (or visible) absorbing analytes in the HPLC column eluate at one or many wavelengths. This critique of those most popular analyte detector modules will be presented in relation to the previously listed specifications of an 'ideal' detector.

Firstly, they easily meet the ideal requirements in terms of signal response time (msec) primarily because of the extremely rapid physical nature of the photoelectrical effect. However, they do not qualify as ideal detectors since they do not give a rapid response to all analytes. Clearly only those analytes which absorb in the UV (or visible) region

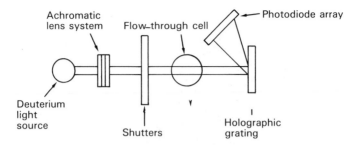

Figure 11. Schematic diagram for a linear diode array photometric detector.

of the electromagnetic spectrum may be detected. This means in practice that certain important groups of compounds such as carbohydrates, inorganics, lipids cannot be detected by a UV detector unless they have first been converted, by pre- or post-column chemical derivatizations to yield chromophoric derivatives.

Secondly, it will now be shown that in terms of sensitivity these detectors fall short of the ideal by several orders of magnitude. Thus we may calculate the sensitivity for a typical modern variable wavelength UV detector, the UVikon 735LC, Kontron Instruments, as follows. The manufacturers have published that short term electrical noise $= 5 \times 10^{-5}$ absorbance units (this corresponds to 2% of the full scale chart deflection on the most sensitive absorbance range of 2.5×10^{-3}). The minimum signal which can therefore be reasonably distinguished and which could arise from a minimum concentration of an UV absorbing analyte (C_{min}) is $2 \times$ noise $= 10 \times 10^{-5}$ $= 10^{-4}$ absorbance units. Assuming that the Beer–Lambert law holds, that the optical path length of the flow-through cells is 1 cm and that the molar extinction coefficient of the UV absorbing analyte is ϵ_{max} then

$$A = \epsilon_{max} \times C_{min} \times l$$

Substituting the above values for A and l we obtain

$$C_{min} = \frac{10^{-4}}{\epsilon_{max} \times 1 \text{ cm}}$$

Now ϵ_{max} has the range $0-10^6$ M^{-1} cm^{-1} but if we assume a mean value for $\epsilon_{max} = 10^3$ M^{-1} cm^{-1}

then

$$C_{min} = \frac{10^{-4}}{10^{-3} \text{ M}^{-1}\text{cm}^{-1} \times 1 \text{ cm}}$$

$$C_{min} = 10^{-7} \text{ Molar.}$$

Further, in order to be able to calculate the sensitivity as defined previously we must assume a value for the molecular mass of the analyte – for purposes of illustration a value of 200 is used. Substituting this value in the last equation we obtain:

$$C_{min} = 200 \times 10^{-7} \text{ g/l}$$

or

$$C_{min} = 2 \times 10^{-8} \text{ g/ml} = \text{sensitivity}$$

This calculated sensitivity value should be compared with that given of $10^{-11}-10^{-12}$ g/ml

Figure 12. Photograph of a UV photometric detector showing the accessibility of its flow-through cell (Kontron 430).

in the second of the specifications for an ideal detector in order to evaluate the UV detector. Clearly for strong UV absorbers with ϵ_{max} greater than 10^3 the sensitivity of the detector improves.

It is difficult theoretically to assess UV detectors with regard to linear working range because of the possibility of non-linear deviations arising at high concentration from non-adherence of the analyte solutions to Beer's Law. Generally speaking however a linear range of order of 10^3 is normally encountered in practice. (See specification iii.)

On the other hand, the UV detector is excellent with regard to the fourth specification regarding the nature and composition of the eluant. Provided that UV transparent solvents are employed—in practice this implies highly purified HPLC quality grade solvents must be used—then the detector behaves in an ideal fashion. This also explains why UV detectors may be used for both isocratic and gradient-elution modes of chromatography. However, UV detectors can be very sensitive to both temperature and flow-rate fluctuations; however, if the elution is operated under isothermal conditions then these effects disappear (13) and the detector conforms to the ideal specification.

The last three ideal specifications are concerned with the flow-through cell and its

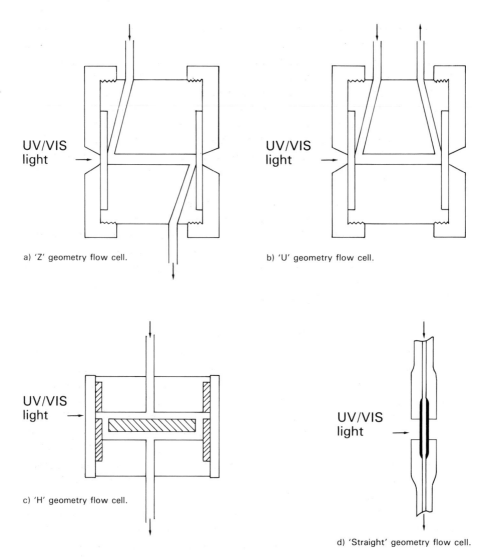

a) 'Z' geometry flow cell.

b) 'U' geometry flow cell.

c) 'H' geometry flow cell.

d) 'Straight' geometry flow cell.

Figure 13. Schematic diagram for the four types of commercial flow-through cells used in conjunction with UV detectors.

coupling to the HPLC column. It is evident to the author that a number of manufacturers have carefully considered the question of coupling the flow-through cells to the column. In particular a number of films now make the flow-through cells readily accessible not only for purposes of coupling to the column, but also so that the cells may be inspected for faults (e.g. air bubbles) and that cells of different sizes may be interchanged as required by the changes in column dimensions, for example see *Figure 12*. It is also clear to the author that most makers of HPLC UV detectors have studied the design of the optical flow-through cells which they employ. Thus in a review of this topic covering 23 manufacturers (14) four different designs were discussed and these are shown in schematic form in *Figure 13*. However, since none of the designs shown in *Figure 13* could have imparted a radial flow to the incoming eluate stream as required by specification number (iv) further cell design work needs to be done. Generally, the flow-through cells reviewed had dimensions for the optical measuring conduit of 1 mm (i.d.) and 10 mm (l). This leads to the measured volume of eluate being 7.8 μl, which in practice is suitable for use with most 4.6 mm (i.d.) analytical HPLC columns. The majority of the commercially available smaller volume cells are made by shortening the cell length but it should be remembered that this will lead to a lowering of the sensitivity of the detector (see sensitivity calculation). Therefore there is still considerable room for development work in the design of the flow-through cells and especially for the low-volume ones required to monitor microbore HPLC (i.d. ≤ 2 mm) columns.

To conclude, this critique of UV detector/flow-through cell combinations has shown that it approximates, in many applications, to an ideal analyte detection module. It is because of this and because of its relative cheapness, ease of use and reliability that it is the most widely used detector in HPLC chromatography. Finally it should perhaps be noted that it is possible for the chromatographer to comprehensively test any fixed wavelength UV-photometric detector by adopting the detailed procedures given in the ASTM standard AN1/ASTM E 685-79 (15).

3.5 **The central control and data output module**

These two modules, numbered in accordance with the block diagram to *Figure 1* will be considered together since they are necessarily closely related. In terms of the development of HPLC systems, the first data output modules were devices, normally potentiometric strip-chart recorders, which received the amplified and transformed electrical response signal from the detector module. In the case of the UV/visible photometric detector, a $0-10$ mV potentiometric was/is commonly employed, with a variable chart speed which enabled it to match the speed of the HPLC separations being performed. This common experimental arrangement which leads to the production of a chromatogram in paper form is still popular and so routine that it will not be discussed further here. Further, it used to be common practice to manually perform all of the requisite calculations from the chromatogram, namely retention times, peak heights, peak areas and the calculation of the various column chromatography parameters such as column efficiency, plate number, capacity factors etc. With the advent of digital computers it then became usual to perform these calculations using an off-line main frame computer (16). However, during the last decade, because of the rapid development of analogue-to-digital converters and of the tremendous reduction in costs of small but powerful digital computers, it has become common practice for the manufacturers of HPLC equipment to offer small dedicated computerized data systems which will perform these calculations. Now whilst it is not proposed to give a detailed discussion of these systems because so many exist, some illustrative examples of the way in which they handle experimental chromatographic data will now be given. Thus in *Figure 14* an experimental chromatogram is shown together with the 'drawn-in' base lines required to perform the peak area calculations. This 'drawing-in' process together with the peak area calculations may be performed routinely with all of the microprocessor chromatographic integrator data systems at present being marketed.

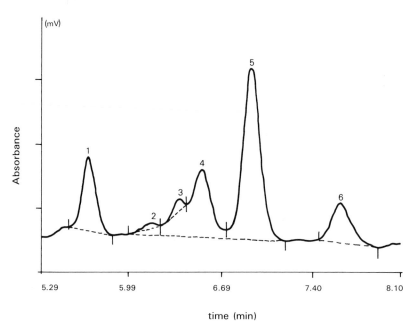

Figure 14. A typical HPLC chromatogram with base lines 'drawn in' (by the chromatographer) in dotted lines.

By contrast *Figure 15* shows how one modern chromatographic data control system (Kontron 450) intelligently samples data for peak area measurements. In *Figure 15* short vertical lines above the chromatogram trace indicates that at this time data points are being stored. As can be seen from a study of *Figure 15*, when the signal is constant only a few data points need to be and are stored but, as soon as the signal starts to change an increasingly high number of data points are stored. This use of a machine intelligent sampling procedure leads to precision peak area measurements and does not overload the computer memory.

A more recent development in the application of modern computer technology to HPLC instruments is to combine the chromatographic data handling system with an automated instrument control system. *Figure 16* shows such a combined instrument control unit (Kontron 450) combined with an intelligent data processing system to form an HPLC 'work station'. When such a control device is interfaced to an HPLC system fitted with an automatic sampling device (*Figure 1*) it is possible to perform chromatography automatically for 24 hours per day, 7 days per week. The major advantages of this new experimental situation are that HPLC separations may be optimized automatically as part of a research programme (17) or alternatively routine analytical work may be completely automated, including quality control and statistics and the production of laboratory reports (18).

The very latest development in the area of computerized laboratory aid devices is that of the versatile 'work station'; that is to use computers which are not dedicated to an HPLC system alone. Such a 'work station' is a device which may be interfaced to a variety of analytical instruments, including HPLC, mass spectrometers, IR spectrometers etc., which will operate them and perform any required calculations on the

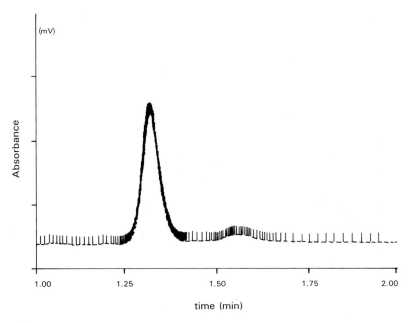

Figure 15. Raw data storage with dynamic data point matching.

generated data and further has powerful graphic display facilities. Further these 'work stations' are able to access large collections of relevant numerical data and can perform library searches on the data. Clearly for those HPLC systems in which say a mass spectrometer is employed as the detector module such 'work stations' are likely to be very attractive to the chromatographer. The major problem with all of these computer devices is that they tend to change the definition of HPLC given in the introduction to this chapter to that of <u>H</u>igh-<u>P</u>riced <u>L</u>iquid <u>C</u>hromatography! The experimental chromatographer can only hope that the capital costs of such equipment will ultimately be reduced so that most laboratories will be able to afford to purchase because there is no doubt that they remove much of the drudgery from HPLC practice.

3.6 Assessment of HPLC instrumentation

Finally, to conclude this section a short account of the experimental method by which an overall assessment of any HPLC system for the separation of small molecules, will be given. The assessment of some of the individual modules has already been discussed, for example, detector flow-through cell modules. However, the key assessment is that of the system operating as a whole since this is the way in which it will be used in practice. The accepted method is that of performing a reported separation of a known test mixture involving a similar HPLC column, eluant system and elution mode and detector, and then comparing the results with those published. For HPLC systems designed to separate small molecules the choice of test mixture presents little difficulty, especially if one is employing commercial HPLC columns since these are normally sold with a test chromatogram (19). Repeating the HPLC separation of the test mixture at regular intervals (e.g. week/month) enables one to check the overall performance

Figure 16. Typical combined HPLC instrument control unit and intelligent data processing unit and output module (Kontron 450).

of the system. If the HPLC separation of the test mixture deteriorates then a number of guides (20) and books (21) exist to help the chromatographer locate and identify the exact cause of the problem. However, as the previous presentation will hopefully have demonstrated, modern HPLC instrumentation for the analysis of mixtures of small molecules is generally well designed, well engineered and robust. Further, since a range of excellent analytical HPLC columns (4−6 mm i.d.) is now widely available (Chapter 1), the separation of mixtures of small molecules using the currently available instrumentation has now become routine practice in many laboratories thanks to the efforts of the instrument designers. As will be shown in the next section, this satisfactory experimental situation has yet to be achieved for the separation of mixtures of large molecules.

4. MODULAR HPLC INSTRUMENTATION FOR THE SEPARATION OF MACROMOLECULES

This section will consist of three parts. In the first part the origin of the experimental problems which arise from the fact that the analyte mixtures consist of macromolecules will be given and discussed. In the second part, a systematic discussion of the modifications required to enable the previously presented HPLC instruments to overcome certain of these new experimental problems will be presented. Finally in the third and final part the section will be concluded with a discussion of some proposed methods of assessment of such HPLC systems designed to separate macromolecules.

4.1 **The origin of the additional problems encountered when separating mixtures of macromolecules by HPLC**

The separation of mixtures of macromolecules, especially if these originate from biological materials as opposed to synthetic mixtures of polymers, present a number of practical problems over and above those encountered when separating mixtures of small molecules by HPLC techniques. In the first place, biological materials are generally exceedingly complex mixtures which contain very large numbers ($10^2 - 10^3$) of compounds some of which vary greatly in molecular size and in concentration (22). By contrast, many of the mixtures of small molecules currently being analysed by HPLC consist of the products of organic synthesis or of pharmaceutical preparations so that the number of components is generally speaking small ($\not> 10$). As a result of this major difference in the complexity of the mixtures encountered it should come as no surprise to the reader to learn that the application of HPLC to the separation of macromolecules is normally only one of many separation procedures which form a complete separation protocol. For example, in a recent study of various overall protocols for the purification of proteins (23), ion-exchange, affinity and size-exclusion HPLC techniques were all employed at various stages of the complete purification procedure, together with older separation techniques such as fractional precipitation. Thus the biochemical chromatographer separating macromolecular mixtures has to gain experience of all of the types of chromatography (see Section 1) and to *design* the overall separation strategy.

Secondly, some of the substances present in these complex biological mixtures may differ in molecular structure by very little indeed; for example, consider a pair of proteins of molecular mass 100 000 in which an iso-leucine residue in one is replaced by a leucine in the other. Now clearly such a pair of proteins would have identical bulk physical properties and yet they may have completely different biological properties, for example immuno-chemical or enzymatic properties. The separation of such mixtures is thus extremely difficult and either columns of very high resolving power or which use the biological difference, such as affinity HPLC columns (Chapter 6) must be employed.

Thirdly, because of their large size macromolecules tend to present severe problems of solubility. Thus, even when they form solutions (or in some cases sols) they are thermodynamically unstable so that they readily precipitate out of the liquid phase if they are subjected to even small changes in say pH, temperature, ionic strength or solvent composition, for example addition of organic solvents. Clearly, since HPLC is by definition chromatography performed in the liquid phase, the macromolecular mixtures must be kept in this phase if the separation is to be successful, that is the analytes must not be allowed to precipitate out of solution either on the column or even off the column.

Fourthly, some biological macromolecules, for example metallo-enzymes, are also very susceptible (easily poisoned) to the presence of very small quantities of metal ion impurities which may be present in the eluant, either at source or which may be leached into the eluant from the stainless steel column housing – or the tubing used to connect the various modules.

Table 3. Synoptic listing and commentary on HPLC procedures and instrumentation modules employed for HPLC separation of mixtures of small and large molecules.

Entry number	Procedure/module	Comment	
		Small molecule	*Large molecule*
1	Design and use suitable sample pre-treatment procedure(s)	Optional	Essential/problem
2	Select an appropriate HPLC column for the particular analyte mixture	Essential	Essential
3	Select an appropriate guard column	Optional	Essential
4	Employ a HPLC column thermostat	Optional	Essential
5	Select a suitable eluant(s)	Essential	Essential/problem
6	Design and filter the eluant(s)	Essential	Essential
7	Employ an accurate, precise and pulse-free pump	Essential	Essential
8	Employ a gradient-forming system	Optional	Essential
9	Employ a suitable sample injector	Essential	Essential
10	Employ a detector of suitable sensitivity	Essential	Essential/problem
11	Employ 316 S.S. connectors i.d. $\leq$ 0.25 mm	Essential	Essential/problem
12	Employ a suitable output device, e.g. strip chart recorder, chromatographic integrator, work station	Essential	Essential
13	Employ fraction collector	Optional	Optional

Fifthly, some macromolecules require the presence of specific small molecules as co-factors or as activators in order to maintain biological activity. If the latter are separated from the former during the course of a HPLC separation it follows that the purified macromolecule is biologically inactive and the chromatographer must remember to replace the missing requisite co-factor under suitable conditions, so as to regenerate the activity.

Sixthly, in the case of proteolytic enzymes, it should be remembered that these are stabilized by the presence of small peptides produced by self-digestion or proteolysis of their precursors. In this case the HPLC separation and subsequent detection and assessment of enzymatic activity should be performed rapidly before proteolysis can occur again.

Finally, whilst the problem of detecting the separated macromolecules may often be achieved using the same detector modules as those employed for small molecules, for example UV/VIS photometric detectors, it should not be forgotten that the problem of the assay of the biological activity remains. Thus for example, most proteins absorb at 280 nm but only the enzymes (which are also proteins) possess enzymatic activity, so that if a protein mixture is believed to contain enzymes then these have to be detected and determined separately. Thus in general most macromolecular HPLC separations require additional detector modules to those discussed in the previous section. In spite of these difficulties many successful separations of biological macromolecules have been performed involving HPLC techniques and some of these were discussed in the main Introduction whilst others form the subject matter of the remaining chapters of this text. That this is the case, implies that the column and instrument manufacturers have successfully modified the HPLC systems so as to overcome the problems listed in paragraphs three to six above and in the following section the design principles behind these modifications will be discussed.

4.2 **HPLC instrumentation for separating mixtures of macromolecules**

Since the HPLC instrumentation required to separate macromolecules consists either of modifications or additions to the combination of the various modules required to successfully separate mixtures of small molecules, this discussion will be based on the latter. In order to aid this discussion a synopsis of the discrete steps in the HPLC separation process for both large and small molecules, together with the corresponding module requirements, are listed in *Table 3*, and these are accompanied by a comment indicating whether or not they are essential or optional and whether problems exist.

A comparative study of the comment column in *Table 3* shows that the HPLC separation of mixtures of large molecules leaves little room for optional modules. Thus entry no. 3 of *Table 3* is concerned with the guard column requirements and, as can be seen, the comment for separations of small molecules is that this is optional. However, since columns suitable for the separation of macromolecules are generally much more expensive than those required for small molecules it is clearly sensible to employ guard columns in order to protect them and hence the appropriate comment on this entry is that it is essential.

Four of the entries shown in the comment column of *Table 3* for macromolecules also indicate that problems exist. The subjects of the first, tenth and thirteenth entries in *Table 3*, namely sample pre-treatment procedures, choice of detector and collection of separated fractions, are all connected in that they arise from the complexity of the analyte mixtures, a subject discussed previously. The subject of the first entry constitutes the major, at present unsolved, problem in the separation of macromolecules. As a result it may be confidently predicted that much of the instrument manufacturers research and development efforts at the present time lies in this area of automated sample pre-treatment devices suitable for mixtures of macromolecules because this area must be automated if the whole HPLC process is to be automated. The problem indicated in the tenth entry of *Table 3*, concerning the choice of detector for macromolecules was also discussed in the previous part. Therefore it is sufficient to note here that because of the lack of a universal, sensitive detector for macromolecules two or more of the available detector molecules, arranged in tandem, may need to be employed. Alternatively if a single detector module of the type already discussed in the previous section is used, then discrete fractions of the eluate must be collected for subsequent 'off-line' analysis by say, gel electrophoresis, immuno- or bio-assay procedures. This alternative practice accounts for the optional entry number 13 in *Table 3* regarding the provision of a fraction collector.

The remaining 'problem' entries in *Table 3* are entry numbers 5 and 11, and those are concerned with the choice of eluant and the use of short pieces of small internal diameter 316 stainless steel tubing to connect the various instrument modules together to form the complete HPLC system. The problem that arises from this practice is that some of the common aqueous salt buffers used as eluants corrode stainless steel and the trace metal ions leached out may react with the macromolecules being separated. Clearly if this corrosion problem exists, then it is also likely to affect all of the other wetted parts of the HPLC system, for example the sample application module, the stainless steel components of the HPLC column, namely the housing and the frit employed at the bottom of the column to retain the support material, and the eluant pump heads if these are made of stainless steel. A number of different approaches have been employed in order to overcome this corrosion problem. First, the wettable stainless steel parts listed above may be replaced by inert substances such as titanium alloys, Kel-F®, Teflon, sapphire, quartz and polyethylene. Such modified HPLC systems are generally called bio-compatible systems since in many cases the bio-macromolecules are adsorbed less on these new surface materials than on stainless steel. In the second approach to the stainless steel corrosion problem, the buffer solutions forming the eluant mixtures are modified by the addition of millimolar amounts of EDTA or EGTA (ethyleneglycol-bis-β-aminoethyl ether-N,N,N^1,N^1-tetraacetic acid) so that the leached out metal ions are converted into inert organo-metallic complexes (24). The third solution to the problem consists of a modification to the eluant delivery module such that the chosen aqueous salt buffers are first placed in leak-proof bottles and then continually sparged with helium. This reduces the oxygen content of the buffers to such an extent that they do not then corrode stainless steel (25). Which one of these three solutions to the problem of corrosion will be generally adopted (if any) by the HPLC instrument manufacturers remains to be seen.

Finally on the subject of instrumentation for HPLC of macromolecules, as stated in the first part of this section, the total separation process is generally likely to involve more than one type of chromatographic column. But many chromatographers separating mixtures of different types of small molecules by HPLC also have to employ different types of columns. Thus the essential instrumentation problems encountered when dealing with macromolecules parallel those faced when separating small molecules, with the exception of those additional instrumental problems listed in *Table 3* and discussed above. It is thus interesting to observe that various HPLC instrument makers have recently started to market specific modular systems designed to separate macromolecules. For example, Applied Biosystems offer a protein and peptide purification system; Bio-Rad offer a HRLM™ system for biomolecules; Dionex offer a Biol LC™ series of systems for the analysis of carbohydrates, proteins, bio-ions in which all of the eluant 'wettable' surfaces are metal-free as also is the auto-sampler and fraction collector!; Pharmacia-LKB offer a FPLC® system; Perkin-Elmer a IsoPure Bioseparation system whilst Waters offer an Advanced Protein Purification System (Waters 650) which may be worked in a cold room! Full details of these and other such systems may be obtained from the manufacturers (see Appendix), abstracted details may be obtained from the series of papers by McNair (7). Generally speaking the design specifications for the separate component modules of these systems are identical to those already presented and discussed in Section 3 and so they will not be considered further.

The final topic to be discussed will be that of the experimental assessment of any such HPLC system for the separation of macromolecules and this forms the subject of the concluding part of this section.

4.3 The assessment of HPLC systems designed to separate macromolecules

The accepted method for the assessment of any HPLC system is to perform a reported separation of a known test mixture involving a similar HPLC column, eluant system and elution mode and detector, and then to compare the results obtained with those published. Such a procedure also enables the chromatographer to assign the column's performance to a numerical scale, normally the number of theoretical plates, by performing the relevant calculation on the chromatograph obtained (see Chapter 1). Further if the test separation is repeated at various intervals of time then the deterioration of the column may be continually assessed in a quantitative fashion. Now as stated at the end of Section 3 the assessment of HPLC columns (and systems) using a test mixture of small molecules presents little problem since the latter is often provided with the column by the manufacturer. No such satisfactory situation yet exists for the assessment of HPLC columns (and systems) for the separation of macromolecules and the reasons for this will now be given. In the first place, the test mixtures should always be similar in size and chemical nature to the actual macromolecules being separated and hence because of the extremely wide range of the latter the provision of a large variety of test mixtures is necessary to cover all eventualities. Secondly it should be remembered that it is not easy to prepare, rigorously purify, to assay the purity and to keep macromolecules pure, especially those of biological interest. Indeed, it is because of this that the biochemist is interested in HPLC techniques in the first place! As a result, at the present time it is general practice for those chromatographers using HPLC methods to separate macromolecules to devise their own test mixtures. For example, for protein work mixtures of the following macromolecules have been used: bovine serum albumin, chymotrypsinogen, cytochrome *c*, immunoglobulin G, insulin, lysozyme, ovalbumin and trypsin. Some laboratories have used the commercially available test mixtures of small molecules to assess the performance of columns which have been used to separate proteins. However, these studies (26) have shown that the calculated theoretical plate numbers indicate only how well the column was packed and not how well it may perform the separation of the proteins.

The problem of interlaboratory assessment of the HPLC columns and systems under discussion therefore still remains to be solved and it is to be hoped that the various manufacturers will address themselves to this problem because of its practical importance. Until this is solved the problem of the overall assessment of HPLC columns and instruments used to separate macromolecules cannot be satisfactorily achieved.

5. REFERENCES

1. Tswett,M.S. (1903) *Proc. Warsaw Soc. Nat. Sci. Biol., Sect. 14*, Minute No. 6.
2. Halpaap,H. (1980) *Kontakte,* **3**, 37.
3. Martin,A.J.P. and Synge,R.L.M. (1941) *J. Biochem.,* **35**, 1358.
4. Snyder,L.R. and Kirkland,J.J. (1974 and 1979) *Introduction to Modern Liquid Chromatography.* 1st and 2nd editions, Wiley-Interscience.
5. Snyder,L.R. and Stadalius,M.A. (1986) *High Performance Liquid Chromatogr.,* **4**, 195.

6. Knox,J.H. (1977) *J. Chromatogr. Sci.,* **15**, 352.
7. McNair,H.M. (1987) *J. Chromatogr. Sci.,* **25**, 564.
8. Oliver,R.W.A. and Basey,A. (1983) *On the Design of Flow-Through Optical Cells in HPLC.* Video – University of Salford.
9. Bakalyar,S.R. and Spruce,B. (1983) *Achieving Accuracy and Precision with Rheodyne Sample Injectors.* Rheodyne Technical Notes No. 5.
10. Maggs,R.J. (1969) *J. Chromatogr. Sci.,* **7**, 145.
11. Scott,R.P.W. (1986) *Liquid Chromatography Detectors.* 2nd edition, Elsevier.
12. George,S. and Martin,R. (1982) *Design Criteria for a Linear Diode Array Detector for Use in Liquid Chromatography.* Hewlett Packard Technical Notes.
13. DuPont Liquid Chromatography Technical Report E-34266 (1980).
14. Basey,A. and Oliver,R.W.A. (1982) *Lab. Practice,* **31**, 553.
15. Annual Book of ASTM Standards (1979) Philadelphia, USA.
16. Oliver,R.W.A. and Sugden,J. (1979) *Chromatographia,* **12**, 620.
17. Bruker Summit – Automated Optimisation System for Liquid Chromatography (1985) Bruker Spectrospin Ltd, pp. 4.
18. Henshall,A. and Schibler,J.A. (1987) *International Laboratory,* **17**, 28.
19. *Merck Spectrum* (1987) **3**, 34.
20. HPLC Troubleshooting Guide No. 826 (1986) Supelco Inc.
21. Walker,J.Q., Jackson,M.T. and Maynard,J.B. (1977) *Chromatographic Systems, Maintenance and Troubleshooting.* Academic Press Inc., London.
22. Oliver,R.W.A. and Oliver,S.A. (1974) *The Analysis of Children's Urine.* Heyden.
23. Bonnerjea,J., Oh,S., Hoare,M. and Dunnill,P. (1986) *Biotechnology,* **4**, 954.
24. Mayhew,J., Ziesk,L.R., Kinsey,W. and Schaefer,J. (1986) *Pittsburgh Conference Proceedings,* 516.
25. Glatz,B., Goetz,H. and Schrenk,W. (1987) Hewlett Packard Publication No. 5.
26. Pfannkoch,E., Lu,K.C. and Regnier,F.E. (1980) *J. Chromatogr. Sci.,* **18**, 430.
27. Burlingame,A. *et al.* (1990) *Proceedings of the 2nd International Conference on Mass Spectrometry Applied to Health Sciences.* San Francisco 1989. In press.

CHAPTER 3

Size-exclusion HPLC of proteins

GJALT W.WELLING and SYTSKE WELLING-WESTER

1. INTRODUCTION

1.1 Principles of SE-HPLC

Since the introduction of Sephadex in 1959 (1) size-exclusion chromatography (also termed gel filtration, gel permeation or molecular sieving) has become exceedingly popular in the field of protein purification. The high-performance mode of this type of liquid chromatography became operational in 1980 after the development of uniform rigid particles of which the uniform pores were sufficiently large to be entered by protein molecules. The rigidity of these chromatographic supports enables high solvent flows to be maintained by the application of high pressure. Separation by size-exclusion (SE)-HPLC depends on differences in the size, more precisely, the hydrodynamic volume of the proteins in a sample. The larger molecules do not enter the pores of the column particles and are excluded so that they are eluted in the void volume of the column (V_0). The pores of a column particle are differentially accessible to smaller particles, depending on their size. This volume of the column is called V_i. The total accessible volume (V_t) is the sum of the volume outside the particles (V_0) and the volume accessible inside the particles: $V_t = V_0 + V_i$ (see *Table 1* for definition of symbols used in this chapter).

In SE-HPLC one buffer is used to elute the proteins. In other modes of HPLC, gradient-elution is generally used and proteins can be separated over the whole gradient volume. In SE-HPLC, separation is possible only between V_0 and V_t and, as a consequence, only a limited number of protein peaks (seldom more than ten) can be separated. In this respect, the support pore volume ratio (V_i/V_0) is very important in determining the resolving power of the columns. A smaller V_0 and a larger V_i means that more of the total column volume can be used for separation. Unfortunately, values for these pore volume ratios are not routinely provided by the manufacturers.

In ideal SE-HPLC there is a linear relationship between the logarithm of the molecular weight (M) and the elution volume (V_e) of a protein. Therefore, SE-HPLC can be used not only to purify a protein but also to determine its molecular weight (with an accuracy of $\sim 10\%$).

The elution volume (V_e) can be described by

$$V_e = V_0 + K_D V_i$$

K_D is the distribution coefficient; when we substitute V_i with $V_t - V_0$ we obtain

$$K_D = \frac{(V_e - V_0)}{(V_t - V_0)}$$

Table 1. SE-HPLC nomenclature.

V_0	— void volume; interparticle volume; volume available to all molecules.
V_i	— pore volume; intraparticle volume; volume that is differentially accessible depending on the size of the protein.
V_e	— elution volume of a protein.
V_t	— elution volume of a small molecule e.g. nucleoside monophosphate, sodium azide, DNP-alanine; total accessible volume.
K_D	— distribution coefficient; fractional pore volume available to a protein molecule.
R_s	— specific resolution of pairs of proteins.
V_i/V_0	— support pore volume ratio.

Table 2. General scheme for pre-SE-HPLC purification.

1. Extraction/homogenization of protein-containing material.
2. Fractionation by ammonium sulphate precipitation.
3. Centrifugation.
4. Desalting by dialysis or conventional SE chromatography followed by further purification, which generally involves one or more chromatographic steps.

NB In the purification of a membrane protein, step 1 involves extraction by a detergent; step 2 is omitted and step 3, centrifugation is directly followed by chromatographic purification.

K_D varies between 0 and 1 and plots of K_D against log M are generally linear between K_D values of $0.15-0.80$ (2). A calibration curve can be constructed not only by plotting log M versus K_D but also against V_e (an example will be given later in Section 2.2 and *Table 4*). The calibration curve provides information on the range of molecular weights that can be covered by a particular column. Furthermore, the smaller the slope of the calibration curve, the better is the separation between proteins of different molecular weights. In addition, the shape of the peak (the peak width) is important although this is not visible in the calibration curve. For example, a steeper calibration curve and narrower peaks may give a better separation. Therefore the resolution between pairs of proteins (R_s) has to be taken into account. A practical example will be given later in Section 2.2.

A generally applicable system for SE-HPLC of proteins consists of one pump that can deliver the eluant mixtures at accurate flow-rates of $0.1-1.0$ ml/min; an injection valve for application of sample volumes up to 200 μl; an optical flow-through cell compatible with a UV-detector with at least 280 nm detection, or if higher sensitivity is required, detection at $210-220$ nm; a chart-recorder and one or two columns that cover the desired molecular weight range and that are compatible with the selected mobile phase. The system is generally operated at room temperature.

1.2 General scheme for purification of complex protein mixtures before utilizing SE-HPLC

The purification of proteins from complex biological mixtures generally involves a number of pre-purification steps before SE-HPLC can be performed and these are listed in *Table 2*.

Essentially the same scheme can be used for recombinant-DNA produced proteins in *Escherichia coli*. These proteins as well as membrane proteins often show a tendency to aggregate which means that detergents (preferably mild non-ionic detergents) have

Table 3. Properties of SE-HPLC column materials suitable for purifications of proteins[a].

Column name	Particle size (μm)	Pore size (nm)	Globular protein fractionation range (kd)	pH stability	V_i/V_0	Manufacturer/ supplier
TSK 2000 SW	10	13	1 – 50	2.5 – 7.5	0.92	1
TSK 3000 SW	10	24	5 – 400	2.5 – 7.5	1.33	1
TSK 4000 SW	13	45	40 – 1000	2.5 – 7.5	1.52	1
Superose 12	10	25	1 – 300	1 – 14	1.83	2
Superose 6	13	40	5 – 5000	1 – 14	2.17	2
Zorbax GF-250	4	15	10 – 250	3.0 – 8.5	1.08	3
Zorbax GF-450	6	30	25 – 800	3.0 – 8.5	1.08	3
Polyol = Si300	5,10	30	10 – 500	2.0 – 8.5	–	4
Polyol = Si500	10	50	40 – 900	2.0 – 8.5	–	4
SynChropak GPC 100	5	10	5 – 200	2.0 – 8.0	1.08	5
SynChropak GPC 300	5	30	10 – 670	2.0 – 8.0	0.95	5
SynChropak GPC 500	7	50	10 – >670	2.0 – 8.0	1.00	5

1, Toyo Soda (Tokyo, Japan), Beckman-Altex (Berkeley, CA), Varian (Walnut Creek, CA), Bio-Rad (Richmond, CA), LKB (Bromma, Sweden), Chrompack (Middelburg, The Netherlands); 2, Pharmacia (Uppsala, Sweden); 3, DuPont (Wilmington, DE); 4, Serva (Heidelberg, FRG); 5, Synchrom (Lafayette, IN).
[a]Larger pore-sizes are available for most of these columns. Other column materials that can be used for SE-HPLC are the I-125 and I-250 material (Waters, Milford, MA), the LiChrospher diol series (Merck, Darmstadt, FRG) and the polymer-based TSK-PW series (Toyo Soda). The SynChropak GPC series are also sold as Aquapore OH (Brownlee, Santa Clara, CA). Most of these column supports are only available as packed columns; the SynChropak GPC series is available as bulk support. The support pore volume ratios for the TSK and Superose supports are from refs 3 and 6, respectively, those for the Zorbax and the SynChropak supports are from the manufacturer.

to be added or denaturing agents have to be used during extraction and subsequent chromatographic steps. It should perhaps be noted that (in a pre-purification process), SE-HPLC may also be included (step 4, *Table 2*) to advantage because of its speed and high recoveries.

2. CHOICE OF COLUMN

2.1 Columns available for SE-HPLC and their properties

SE-HPLC column materials generally are chemically modified silicas, which limits their use to a pH range of ~2 – 8. For example, SynChropak is a glycerol propyl-type bonded silica; TSK-SW types are glycol ether-type bonded silica of which the surface is covered with hydroxyl groups; the Zorbax Bio Series GF columns contain diol-bonded silica particles with a metal (zirconium)-oxide-stabilized surface. Other column supports are TSK-PW, which is a hydroxylated polyether copolymer and Superose, consisting of agarose, cross-linked with a mixture of long-chain di- and poly-functional epoxides, followed by further cross-linking with short-chain bifunctional cross-linkers (3). Since the latter two column supports are not based on silica, they are stable over a wider pH range.

Fractionation limits are of course directly related to the pore size of the column particles but other parameters are also important in the separation process. The resolution is proportional to the molecular weight and will decrease especially for higher molecular

Table 4. Experimental procedure for the construction of a calibration curve for a SE-HPLC column.

1.	Add 2 mg of SDS to 100 μl of water containing BSA, ovalbumin and trypsin inhibitor (50 μg each) in a microcentrifuge tube[a].
	NB Take twice the amount of SDS for hydrophobic protein samples.
2.	Keep the tube in a boiling waterbath for 3 min.
3.	Inject 100 μl into the SE-HPLC system containing the column of choice.
4.	Monitor the absorbance at 280 nm (0.1 or 0.2 AUFS).
5.	Plot the elution volume (V_e) of each peak against the logarithm of the mol. wt on semi-log paper. The mol. wts (in kd) of the proteins we used for the construction of the curve are: BSA 68; ovalbumin, 43; trypsin-inhibitor, 20; in addition we used the tetramer of the Sendai virus HN protein (272 kd).

[a]Prepare a stock solution containing 5 mg of each protein in 10 ml water; 100−400 μl portions of this solution are stored at −20°C. This reference solution can also be used to check column performance.

weight proteins at higher flow-rates because of slower mass transfer. The viscosity of the sample will also play a role and, as a rule, the viscosity of the sample should not be more than twice as high as that of the mobile phase (4). Resolution is also affected by the sample volume and a maximum of 1−2% of the column bed volume is recommended (5). This is 260−520 μl for a TSK 3000 SW column of 600 mm length and 7.5 mm internal diameter. We generally apply not more than 200 μl to this column. To maintain resolution, the maximum mass load of such a column containing 10 g of silica amounts to 1 mg (5).

A selection of supports suitable for protein purification by SE-HPLC is summarized in *Table 3*.

2.2 Construction of a calibration curve for any SE-HPLC column and calculation of its specific resolution: importance of particle size, pore size and column length

A calibration curve is necessary when the molecular weight of a protein is to be determined by SE-HPLC. The calibration curve is also a suitable tool in the selection of a column for a particular separation. A practical example will now be given which allows the comparison of three small SE-HPLC Polyol columns (100 mm × 4.6 mm i.d.) with particle sizes and pore sizes, respectively: (a) 5 μm, 50 nm; (b) 5 μm, 30 nm; (c) 10 μm, 30 nm. At the same time, this example also illustrates the preparation of sodium dodecyl sulphate (SDS)-treated samples for SE-HPLC. To obtain the calibration curve for each column, the procedure detailed in *Table 4* was followed.

The resulting plots for these three columns are shown in *Figure 1*. The results illustrate the importance of pore size and particle size. It will be remembered from the previous discussion that the smaller the slope of the calibration curve the better is the separation between proteins of different molecular weights. A comparison of the curves given by the two columns (a and b) packed with 5 μm particles that differ only in pore size, 50 and 30 nm, shows that with the 50 nm column a slightly better separation is obtained in the lower molecular weight region (<43 kd) and a much better separation in the higher molecular weight region. This indicates the necessity of a larger pore size for optimal separation of the large SDS−protein complexes. A comparison of the slopes of the curves obtained with columns b and c, differing only in particle size 5 and 10 μm, respectively, suggests that a better separation is obtained with the 10 μm

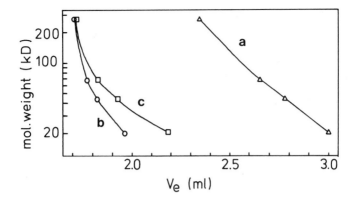

Figure 1. Calibration curves for three Polyol columns. Calibration curves of three SE-HPLC columns (100 mm × 4.6 mm i.d.) (Serva, Heidelberg, FRG). Particle size and pore size respectively were (a) 5 μm, 50 nm; (b) 5 μm, 30 nm; (c) 10 μm, 30 nm. The flow-rate was 0.1 ml/min and 10 μl samples of the tetramer of Sendai virus protein HN (272 kd), BSA (68 kd), ovalbumin (43 kd) and trypsin-inhibitor (20 kd) were applied to the column.

Table 5. Specific resolution (R_s) for pairs of proteins on different Polyol columns.

Protein pair	R_s		
	Column (a) 5 μm, 50 nm	(b) 5 μm, 30 nm	(c) 10 μm, 30 nm
Sendai (HN)$_4$−BSA	3.79	1.77	1.98
BSA−Ovalbumin	3.11	1.70	1.59
Ovalbumin−Trypsin inhibitor	2.82	2.74	2.97

material. However, the construction of calibration curves alone is not always sufficient for proper selection of columns. Smaller column particles generally result in narrower peaks and therefore the two 30 nm columns (b and c) may still result in a similar separation. To obtain a more complete picture of the column performance, the specific resolution R_s between pairs of proteins on each column can be calculated using the following general equation.

$$R_s = 2(V_{e2} - V_{e1})/[(V_{p2} + V_{p1}) \times (\log M_1/M_2)]$$

where V_e, V_p and M are the elution volumes, peak volumes and the molecular weights, respectively. The peak volume is determined as 1.7 × peak-width at half-height (7). As an example, the specific resolution between ovalbumin and trypsin inhibitor on the 50 nm column will now be calculated. The elution volumes (V_{e1} and V_{e2}) and the peak widths were determined in ml at a suitable chartspeed of the recorder. Multiplication of the latter by 1.7 gives the peak volumes (V_{p1} and V_{p2}). The measured values are then substituted into the equation for the specific resolution as shown below to yield a value for the R_s of 2.82.

$$R_s = 2(3.015 - 2.780)/[1.7 (0.150 + 0.145) \times \log 43000/20000]$$

The corresponding R_s values for the other pairs of proteins and the other two columns are shown in *Table 5*. When we compare the two 30 nm columns, a study of *Table 5* shows that the resolution between bovine serum albumin (BSA) and ovalbumin is slightly higher (1.70 versus 1.59) on the 5 μm column despite the steeper calibration curve. The specific resolution of Sendai $(HN)_4$ – BSA, and BSA – ovalbumin on the 50 nm column is 3.79 and 3.11, respectively, and is higher than that found on the 30 nm columns.

Finally on the subject of the choice of columns it should be noted that the peak-width is proportional to the square root of the column length (5) so that an increase in the column length by, for example, coupling columns will therefore increase the specific resolution.

3. ELUTION SYSTEMS

Two basically different mobile phase systems can be used in SE-HPLC: (i) physiological buffers of (near) neutral pH, sometimes with additives as mild non-denaturing detergents and (ii) SDS-containing buffers and denaturing conditions for sample treatment. In the latter case, the actual size of the denatured protein has to be taken into account when selecting a column. SDS – protein complexes are 2.4 times as large as the original protein (8). An example of a separation of a reference protein mixture subjected to conditions given in the figure caption is shown in *Figure 2*. After using SDS, ion-exchange

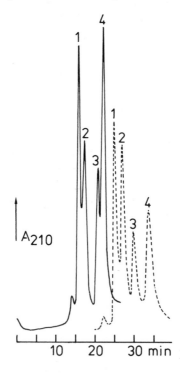

Figure 2. SE-HPLC separations of standard proteins using two different eluant mixtures. SE-HPLC of reference proteins (**1**, BSA; **2**, ovalbumin; **3**, trypsin-inhibitor; **4**, bovine ribonuclease; 2.5 μg of each protein) on a TSK 3000 SW column (500 mm × 7.5 mm i.d.). The column was eluted with 0.1% SDS in 50 mM sodium phosphate, pH 6.5 (———) or with phosphate-buffered saline (PBS), pH 7.2, without SDS (----). For chromatography in 0.1% SDS samples were pretreated as in *Table 4*. The flow-rate was 0.5 ml/min and the absorbance was monitored at 210 nm.

Table 6. List of elution buffers used for SE-HPLC of hydrophobic proteins.

Protein(s)	Elution buffer	Reference
Tick-borne encephalitis virus, Sendai virus, equine infectious anemia virus (EIAV), PDGF receptor	0.1% SDS in 0.05 M sodium phosphate, pH 6.5 or 0.1% SDS in 0.1 M sodium phosphate, pH 7.0	10,11,12,13,14
EIAV, PDGF-fragments	0.02 M sodium phosphate, pH 6.5 with 6 M guanidinium–HCl (GuHCl) 0.01 M sodium phosphate, pH 7.0 with 6 M GuHCl	13,14
Halobacterium halobium	0.1% SDS in 0.2 M Tris–HCl, pH 8 or 0.1 M Tris–HCl, 0.1 M NaCl, pH 7	15
Sendai virus	0.1% decylpolyethyleneglycol-300 in 0.05 M sodium phosphate, pH 6.5	16
Bacteriorhodopsin	0.1 to 5% Triton X-100 in 0.1 M Tris–acetate	17
ATPase	0.1% of the following detergents: SDS, Triton X-100 $C_{12}E_8$, sodium deoxycholate, myristoylglycerophosphocholine in 0.02 M Mops, pH 7, 0.05 M NaCl	18
Blood platelet membrane proteins	0.5% Berol 185 in 0.1 M triethanolamine HCl, pH 7.4	19
E.coli cytochromes	0.05% Sarkosyl, 0.6 M NaCl, 0.01 M Tris–HCl, pH 7.5	20
Ia antigens	0.2% Triton X-100, 0.1% triethylamine, pH 3.0	21
Plasma membrane proteins	01% SDS in 0.155 M NaCl with 8 mM sodium phosphate, pH 7.1	22
Influenza virus	0.1% SDS, Brij 35 or Lutensol ON 70D in 0.1 M sodium phosphate, pH 7	23
β-Glucocerebrosidase	0.1 M sodium citrate, pH 5.0, 50% ethyleneglycol	24
Plasma membrane proteins	0.05% CHAPS in 0.01 M Tris–HCl, pH 7.1, 0.155 M NaCl	25
Human erythrocyte ghosts	0.21% SDS, 0.25% CHAPS, 0.08% reduced Triton X-100 or 0.003% Tween-20 in 0.1 M sodium phosphate, pH 6.5, 0.1 M NaCl	26
Membrane glycoprotein antigen	0.25% sodium deoxycholate in 0.01 M sodium phosphate, pH 7.4, 0.15 M NaCl	27
Bovine rhodopsin	0.88% octylglucoside in 0.06 M sodium phosphate, pH 6.5, 0.15 M NaCl	28
Lymphotoxins	0.1% polyethyleneglycol in 0.1 M sodium phosphate, pH 7.4	29
sn-1,2-Diacylglycerol kinase	2-propanol/heptane/water/TFA, 67/16.5/16.5/0.05	30
Sendai virus	45% acetonitrile in 0.1% HCl	31
Muscarinic acetylcholine receptor	0.1% digitonin in 0.2 M sodium phosphate, pH 7.0, 0.01 M carbamylcholine	32

or reversed-phase HPLC is only possible when SDS is removed (see Section 5). At low ionic strength silica-based SE-HPLC packings act as weak cation-exchangers. They still contain negatively charged silanol groups that have not been modified. To minimize these ionic interactions it is recommended to use salt concentrations above 0.1 M. However high salt concentrations will favour hydrophobic interactions between the protein and the column ligands. This can be useful in particular separation problems but deviations from the linear relationship between the elution volume and the logarithm of the molecular weight of the proteins will occur. Therefore a neutral salt (e.g. NaCl) concentration between 0.1 and 0.5 M is recommended for true SE-HPLC. When SE-HPLC is performed with SDS in the eluant, the ionic interactions are less prominent and salt concentrations between 0.05 and 0.15 M are recommended (9). For SE-HPLC on agarose-based Superose columns 0.15 M salt has proved to be sufficient to suppress ionic interactions between residual carboxylic groups of the support and positively charged solutes (3).

In everyday separation problems we often have to deal with proteins that behave in a more complex manner than the easily soluble reference proteins. Thus, many proteins tend to aggregate and detergents or denaturants have to be included in the mobile phase. *Table 6* lists a number of buffers that have been used for SE-HPLC purification of a number of large hydrophobic proteins and study of this table shows that 0.1% SDS in a sodium phosphate buffer has frequently been used. SDS may denature the protein but as we will see later in Section 4.1, proteins subjected to these conditions are still useful in immunological studies. Other commonly used additives are mild nonionic detergents which result in broader peaks but which are compatible with further chromatography on ion-exchange or reversed-phase columns. In addition, organic solvents can be used to overcome aggregation but peaks are generally broad and, depending on the protein, precipitation may occur at high organic solvent concentrations.

4. ANALYSIS OF ELUATE FRACTIONS

Because of the complexity of the original biological matrix starting material it cannot be assumed that the separate UV absorbing peaks used to monitor the SE-HPLC are in fact due to a single protein. Further analysis of the eluate fractions must therefore be performed preferably by several techniques, the exact choice of which depends primarily on the purpose of the purification.

4.1 **SDS—polyacrylamide gel electrophoresis**

SDS—polyacrylamide gel electrophoresis (SDS—PAGE) can be performed in order to estimate molecular weights of the proteins in the UV absorbing fractions and in combination with the highly sensitive silver staining technique it can be used as a criterion for purity. Unfortunately, electrophoresis on slab gels of 9 × 13 cm takes 5—6 h and staining is generally carried out the next day after fixation and washing in methanol. More rapid mini SDS—PAGE systems are available now from LKB (Bromma, Sweden), Bio-Rad (Richmond,CA, USA) and Hoefer (San Francisco, CA, USA). The size of the mini-gels is about 7 × 8 cm and they take a few hours to run. Staining can be done the same day. Even more rapid is the PhastSystem (Pharmacia, Uppsala, Sweden). An SDS-gel of 3.8 × 3.8 cm takes 30 min to run. Coomassie and silver staining take

UNIVERSITY OF WOLVERHAMPTON
Harrison Learning Centre

ITEMS ISSUED:

Customer ID: WPP60969024

Title: Gel electrophoresis : proteins
ID: 7608368708
Due: 30/10/12 23:59

Title: HPLC of macromolecules : a practical
approach
ID: 7608137404
Due: 30/10/12 23:59

Total items: 2
Total fines: £2.80
09/10/2012 14:20
Issued: 7
Overdue: 0

Thank you for using Self Service.
Please keep your receipt.

Overdue books are fined at 40p per day for
week loans, 10p per day for long loans.

30 and 60 min respectively, with both the electrophoresis and staining processes being fully automated. One difference between these systems which has to be taken into account is the maximum volume of the sample that can be applied to the gel. These volumes are for the large gel, the mini-gel and the PhastSystem-gel 80, 30 and 1 μl respectively.

4.2 Immunological activity

When the protein is purified by SE-HPLC, eluate fractions may be used as antigens in immunological test systems. An enzyme-linked immunosorbent assay (ELISA) can be used to determine the remaining immunological activity after chromatography. Well-defined monoclonal or polyclonal antibodies against proteins in a mixture may be useful in an ELISA to establish the identity and purity of proteins in the eluate fractions. Two examples of their use to check SE-HPLC protein eluate fractions will now be given. In the first of these, the starting protein mixture is provided by Sendai virus, a paramyxo-virus of mice. The protein mixture was first extracted from the latter according to the methods detailed by us elsewhere (12). Prior to chromatography the extracted proteins were boiled for 3 min in 4% SDS. The proteins were then separated by SE-HPLC using 0.1% SDS in 50 mM sodium phosphate, pH 6.5 as the mobile phase. After chromatography, the separated proteins still reacted with polyclonal antibodies to the intact virus (antiserum from infected mice) in an ELISA (12). In the second example,

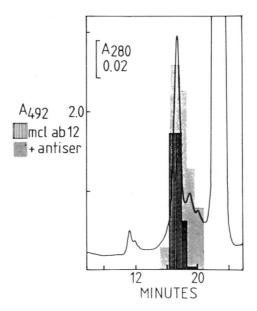

Figure 3. Combined, direct and ELISA product, photometric monitoring of fractions of the SE-HPLC separation of an extract of a viral protein mixture. SE-HPLC of a Triton X-100 extract of Newcastle disease virus strain LaSota. The extract was prepared as described in ref. 12. The TSK 4000 SW column was eluted with 50 mM sodium phosphate, pH 6.5, containing 0.1% SDS. The flow-rate was 1 ml/min and the absorbance was monitored at 280 nm. Fractions (1 ml) were analysed with an ELISA (see *Table 7*). Shaded areas, which represent absorbance at 492 nm, indicate immunological activity. A positive polyclonal antiserum was used as well as a monoclonal antiserum directed against the haemagglutinin-neuraminidase protein (Courtesy of Dr Y. Nagai, Nagoya University School of Medicine, Nagoya, Japan).

Table 7. General scheme for the ELISA of SE-HPLC fractions.

1.	Apply $100-500$ μg of viral protein to the column and elute at 1 ml/min with 0.1% SDS in 50 mM sodium phosphate, pH 6.5
2.	Collect 1 ml eluate fractions in 70×11 mm minisorp-tubes (Nunc, Roskilde, Denmark).
3.	Cover the tubes with a square piece of dialysis tubing, close the minisorp-tube by gently pressing 1.5 cm length of silicone tubing over it.
4.	Dialyse against water overnight in the cold.
5.	Freeze the fraction in the tubes in liquid nitrogen and freeze-dry.
6.	Add 600 μl of 50 mM sodium carbonate, pH 9.6 to each tube and gently mix the contents.
7.	Coat the wells of polystyrene microtitreplates (Dynatech Denkendorf, FRG) with 100 μl (maximally 10 μg protein[a]) from each tube. An approximate measure of the amount of protein is obtained by comparing the absorbance of the peaks with that of a reference protein mixture (see below, Section 6). Coat wells for reaction with a positive and a negative serum. Coat control wells to test non-specific reaction between protein and conjugate. Coat control wells with coating buffer instead of protein. Coating is done in duplicate.

Example	Coat	Add	Result OD_{492}
	Eluate fraction	positive serum and conjugate	+
	Eluate fraction	negative serum and conjugate	−
	Eluate fraction	dilution buffer and conjugate	−
	Coating buffer	positive serum and conjugate	−

8.	Keep the plates for 2 h at 37°C or overnight at 4°C.
9.	Wash the plates three times (5 min) with PBS containing 0.3% Tween 20, 0.2 M NaCl and 100 mg SDS/litre.
10.	Add 100 μl of a suitable dilution (generally 1:100) of a positive antiserum (from mice) in the same buffer supplemented with 0.5% BSA (dilution buffer) to each well and negative serum to control wells (see example). Add 100 μl of dilution buffer to the appropriate control wells (see example).
11.	Incubate for 1 h at room temperature.
12.	Repeat step 9.
13.	Add 100 μl of a suitable dilution of a conjugate (sheep anti-mouse IgG conjugated to horseradish peroxidase) in the same buffer supplemented with 0.5% BSA.
14.	Incubate for 1 h at 37°C.
15.	Add 100 μl to each well of a solution containing 0.2 mg/ml orthophenylene diamine dihydrochloride (Eastman Kodak, Rochester, USA) and 0.006% H_2O_2 in 0.05 M sodium−potassium phosphate, pH 5.6 to visualize the enzyme reaction.
16.	Incubate for 30 min at room temperature.
17.	Add 50 μl of 2 M H_2SO_4 to terminate the peroxidase reaction and read the optical density at 492 nm in a microplate photometer. The optical density is a measure of the immunological activity in each fraction.

[a]Small amounts of protein (less than 1 μg, depending on the protein) may give OD_{492} values of less than 0.2, which is considered as negative.
[b]These conditions may vary for different proteins; they should give an OD_{492} of 2.0 or higher for the positive serum and a maximum OD_{492} of 0.2 for the negative serum.

a detergent extract of Newcastle disease virus (NDV) was subjected to the same SE-HPLC procedure and the eluate fractions were investigated in an ELISA with a positive polyclonal antiserum and a monoclonal antibody directed against the haem-agglutinin-neuraminidase protein of NDV. The photometric results are shown in *Figure 3*. The complete ELISA procedure is detailed in *Table 7*.

Table 8. General scheme for the analysis of immune precipitates.

1.	Dissolve the antigen in PBS, pH 7.2, (0.25 mg/ml).
2.	Incubate 100 μl of antigen solution and 100 μl of antiserum for 30 min at 37°C in a microcentrifuge tube.
3.	Centrifuge in an Eppendorf centrifuge.
4.	Wash the precipitate at least twice with cold PBS.
5.	Add 50 μl of 2% SDS in 10 mM sodium phosphate, pH 7.2.
6.	Keep the tube for 2 min in a boiling water bath.
7.	Inject the solution which contains solubilized IgG and antigen into an isocratic HPLC system equipped with a column on which IgG and the antigen can be separated (see *Table 3*) and elute with 0.1% SDS in 50 mM sodium phosphate, pH 6.5 at a flow-rate of 0.5 ml/min.
8.	Detect IgG at 210 nm and the smaller protein antigens by increasing the sensitivity 16× after elution of the IgG. See *Figure 4*.
9.	Analyse the fractions by SDS−PAGE.

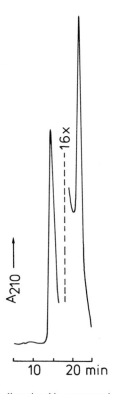

Figure 4. SE-HPLC chromatogram of a dissociated immune precipitate. SE-HPLC of a dissociated immune precipitate (see *Table 8*) consisting of ribonuclease and IgG directed against ribonuclease on a TSK 3000 SW column (500 mm × 7.5 mm i.d.) which was eluted with 0.1% SDS in 50 mM sodium phosphate, pH 6.5 at a flow-rate of 0.5 ml/min. The absorbance was monitored at 210 nm. The sensitivity was increased 16-fold after elution of the IgG in order to detect the antigen ribonuclease.

4.3 Enzymatic activity

In those instances where physiological buffers can be used for elution, the speed of SE-HPLC can result in the purification of a fully active enzyme with almost quantitative recovery. The actual yield will of course depend on the stability of the enzyme. Thus Pfannkoch *et al.* (33) reported recoveries of trypsin activity of more than 86% on different SE columns. Trypsin is a relatively stable enzyme and the purification of labile enzymes may require that the chromatography is carried out in the cold and at the highest possible flow-rate.

5. OTHER SELECTED APPLICATIONS

Tempst *et al.* (14) utilized SE-HPLC in a microscale structure analysis of the platelet-derived growth factor (PDGF) receptor which is an intrinsic membrane protein of 185 kd. Affinity column fractions containing PDGF receptor were made 2% in SDS and incubated at 37°C for 15 min. They were eluted from a TSK 4000 SW column with 0.1% SDS in 0.1 M sodium phosphate, pH 7.0. Fractions were analysed by SDS−PAGE followed by silver staining and, upon rechromatography of one of the fractions, a pure protein was obtained (150 μg of protein). A 50 μg amount of the protein was reduced and carboxyamidomethylated in a buffer containing 2% SDS and the resultant mixture was then extensively dialysed against 0.1% SDS. The protein was dissolved in 2 M urea and formic acid was added (final concentration 70%) followed by 1 mg cyanogen bromide. The resulting fragments were separated on a TSK 2000 SW column that was eluted with 10 mM sodium phosphate, pH 7.0 containing 6 M guanidine hydrochloride. The fragments were subjected to reversed-phase HPLC and further primary structure studies.

Immune precipitates containing IgG (mol. wt 150 kd) and protein sufficiently different in molecular weight can be analysed by SE-HPLC. The analysis of a ribonuclease (mol. wt 13.7 kd)−antiribonuclease immune precipitate will be shown as an example (*Table 8* and *Figure 4*). The lower limit of protein that could be detected in this way was about 25 ng.

6. COLUMN MAINTENANCE

Guard columns are recommended for SE-HPLC columns since these columns are relatively expensive. Their major function is to prevent any dirt or precipitates from entering the column. If the backpressure goes up, the guard column should be eluted in reversed direction and the column frits should be replaced or cleaned by sonication in 3 M HNO_3.

It is recommended to check column performance regularly by chromatography of a reference sample. We use a mixture of BSA, ovalbumin and trypsin inhibitor (50 μg each) at 0.1 or 0.2 AUFS (absorbance units full scale) depending on the column. If performance decreases, the guard column and the analytical column should be checked for voids. If these are found then the columns should be filled with dry column support material. After chromatography, at least several volumes of water should be pumped through the system. Most of the mobile phases used in SE-HPLC contain a considerable amount of salt which will cause corrosion of the metal parts of the HPLC system. If the column will be used again the next day, an alternative is elution overnight with the mobile phase at a flow-rate of 0.05−0.15 ml/min. For longer storage the column is kept in 50% methanol or 0.1% sodium azide.

7. REFERENCES

1. Porath,J. and Flodin,P. (1959) *Nature,* **183**, 1657.
2. Regnier,F.E. (1983) *Science,* **222**, 245.
3. Andersson,T., Carlsson,M., Hagel,L., Pernemalm,P.-A. and Janson,J.-C. (1985) *J. Chromatogr.,* **326**, 33.
4. Barth,H.G. (1980) *J. Chromatogr. Sci.,* **18**, 409.
5. Unger,K. (1984) In Jakoby,W.B. (ed.) *Methods in Enzymology,* Vol. 104. Academic Press Inc., London, p. 154.
6. Alfredson,T.V., Wehr,C.T., Tallman,L. and Klink,F. (1982) *J. Liquid Chrom.,* **5**, 489.
7. Van der Zee,R. and Welling,G.W. (1985) *J. Chromatogr.,* **325**, 187.
8. Reynolds,J.A. and Tanford,C. (1970) *Proc. Natl. Acad. Sci. USA,* **66**, 1002.
9. Takagi,T., Takeda,K. and Okuno,T. (1981) *J. Chromatogr.,* **208**, 201.
10. Winkler,G., Heinz,F.X. and Kunz,C. (1984) *J. Chromatogr.,* **297**, 63.
11. Winkler,G., Heinz,F.X., Guirakhoo,F. and Kunz,C. (1985) *J. Chromatogr.,* **326**, 113.
12. Welling,G.W., Nijmeijer,J.R.J., Van der Zee,R., Groen,G., Wilterdink,J.B. and Welling-Wester,S. (1984) *J. Chromatogr.,* **297**, 101.
13. Montelaro,R.C., West,M. and Issel,C.J. (1981) *Anal. Biochem.,* **114**, 398.
14. Tempst,P., Woo,D.D.-L., Teplow,D.B., Aebersold,R., Hood,L. and Kent,S.B.H. (1986) *J. Chromatogr.,* **359**, 403.
15. Konishi,T. (1982) In Packer,L. (ed.) *Methods in Enzymology,* Vol. 88. Academic Press Inc., London, p. 202.
16. Welling,G.W., Slopsema,K. and Welling-Wester,S. (1986) *J. Chromatogr.,* **359**, 307.
17. Pabst,R., Nawroth,T. and Dose,K. (1984) *J. Chromatogr.,* **285**, 333.
18. Lüdi,H. and Hasselbach,W. (1984) *J.. Chromatogr.,* **297**, 111.
19. McGregor,J.L., Clezardin,P., Manach,M., Gronlund,S. and Dechavanne,M. (1985) *J. Chromatogr.,* **326**, 179.
20. Kita,K., Murakami,H., Oya,H. and Anraka,Y. (1985) *Biochem. Int.,* **10**, 319.
21. McKean,D.J. and Bell,M. (1982) *Protides Biol. Fluids,* **30**, 709.
22. Josić,D., Baumann,H. and Reutter,W. (1984) *Anal. Biochem.,* **142**, 473.
23. Calam,D.H. and Davidson,J. (1984) *J. Chromatogr.,* **296**, 285.
24. Murray,G.J., Youle,R.J., Gandy,S.E., Zirzow,G.C. and Barranger,J.A. (1985) *Anal. Biochem.,* **147**, 301.
25. Josić,D., Hofmann,W., Wieland,B., Nuck,R. and Reutter,W. (1986) *J. Chromatogr.,* **359**, 315.
26. Matson,R.S. and Goheen,S.C. (1986) *J. Chromatogr.,* **359**, 285.
27. Lambotte,P., Van Snick,J. and Boon,T. (1984) *J. Chromatogr.,* **297**, 139.
28. DeLucas,L.J. and Muccio,D.D. (1984) *J. Chromatogr.,* **296**, 121.
29. Fuhrer,J.P. and Evans,C.H. (1982) *J. Chromatogr.,* **248**, 427.
30. Loomis,C.R., Walsh,J.P. and Bell,R.M. (1985) *J. Biol. Chem.,* **260**, 4091.
31. Welling,G.W., Groen,G., Slopsema,K. and Welling-Wester,S. (1985) *J. Chromatogr.,* **326**, 173.
32. Haga,K. and Haga,T. (1985) *J. Biol. Chem.,* **260**, 7927.
33. Pfannkoch,E., Lu,K.C., Regnier,F.E. and Barth,H.G. (1980) *J. Chromatogr. Sci.,* **18**, 430.

CHAPTER 4

Ion-exchange chromatography of proteins and peptides

MICHAEL P.HENRY

1. INTRODUCTION

1.1 Characteristics

High-performance ion-exchange chromatography (HPIEC) of proteins and peptides is a technique that has the following characteristics.

(i)   Instrument based and automated (usually microprocessor controlled).

(ii)   Requires high pressures (usually > 100 p.s.i.).

(iii) Uses microparticulate (< 20 micron) rigid chromatographic media.

(iv)   Applicable to any scale of operation (nanogram to kilogram quantities).

(v)   Highly resolving and rapid (runs are less than 1 h, typically).

(vi)   Capable of high mass yields and activity recoveries (up to 100%).

(vii) Expensive, requires $3-6$ months training.

The phenomena that occur during protein HPIEC are no different from those occurring in the earliest forms of ion-exchange chromatography of these biopolymers. The first ion-exchangers were particulate, large diameter (> 50 micron) porous, soft gels derived from polystyrene, polysaccharides and polyacrylates. Modern high-performance media are microparticulate in nature and are derived from silica, hydroxylated polyethers, polystyrenes and cross-linked, non-porous polyacrylates. The buffers and pH ranges that are employed in HPIEC are the same as those used in soft gel chromatography.

On the other hand, the physical ruggedness of modern ion-exchange packing materials has changed certain aspects of the methodology of this technique. For example, there is less need for low temperatures and the stability they bring, since chromatography is so rapid. Columns may be used many times (several hundred) before discarding. Methods development and in-process monitoring may be accomplished very rapidly, allowing true optimization of a given separation to be achieved. Re-equilibration, clean-up and sterilization are accomplished rapidly without unpacking the column. Closer attention must be paid to the operation and maintenance of the instruments in HPIEC, since they are subject to higher stresses and potential corrosion problems.

1.2 General principles

HPIEC of proteins and peptides is a technique that requires the use of a packed column of particulate material consisting of a rigid, polymeric substrate. To this are more or less permanently bound charged groups (fixed charges) to which proteins and peptides

91

(mobile ions) can be attached via largely ionic interactions. A mobile phase consisting of a buffer solution at a known pH and ionic strength, is pumped through the column till the eluant properties match those of the buffer. The sample is injected onto the ion-exchange packing (bonded phase) under conditions which promote the reversible, non-denaturing binding of the protein or peptide of interest. The biopolymer is then selectively removed (desorbed) from the ion-exchanger by the passage of a mobile phase of the appropriate pH, ionic strength and buffer species. The technique generally requires the employment of a pH and/or ionic strength gradient, to achieve the best resolution in a practical time period.

1.3 Separation mechanism

Modern theories of the ion-exchange process have been proposed (1,2). Briefly the steps in this process can be described as follows.

(i)   In the equilibrated ion-exchanger, prior to sample injection, the fixed charges will be distributed over the surface, perhaps $0.5-1$ nm apart. These groups will be solvated to some degree and will have attracted a buffer ion or ions of the opposite charge.

(ii)   Amongst the fixed charges may be regions of varying degrees of hydrophobicity, depending on the other functional groups present.

(iii)   The mobile ion (protein or peptide) having a region of charge on its surface opposite in sign to the fixed charge, will bind, thereby displacing several buffer ions and solvent molecules. It has been shown (2) that a given protein will bind to an ion-exchanger via a specific region of the surface of the former, and at several points on the latter. This region on the biopolymer will normally bear the highest charge density in the molecule, which may or may not be the same sign or magnitude as the charge on the protein as a whole.

(iv)   The protein or polypeptide will remain bound to the fixed charges of the ion-exchanger until competition from the buffer ions of sufficiently high ionic strength, will displace the biopolymer from its multiple anchorage points. Varying pH may alter the state of ionization at the binding surface of the protein and the ion-exchanger. A change in the buffer ion type may change its displacing power, and also the nature of the solvation of the surfaces of the biopolymer and ion-exchanger.

(v)   As the protein moves into the bulk of the mobile phase, it must create a cavity for itself by breaking hydrogen and other bonds among solvent species. The higher the surface tension of the medium, which normally increases with ionic strength, the more difficult it is for the polymer to create this cavity and desorb from the surface.

(vi)   When the pH, ionic strength, buffer ion type and surface tension are optimal, the protein will elute from the column. The extent to which this happens depends upon the influence of the above parameters (pH, ionic strength, ion type) upon the distribution coefficient of the protein between stationary and mobile phases. In isocratic chromatography this coefficient is finite and constant. In gradient-elution this coefficient is ultimately very large and rapidly varies, giving fast-moving, self-sharpening bands.

(vii) The mobile phase parameters that correspond to protein elution will depend upon the nature of the biopolymer and the ion-exchange surface. Manipulation of these parameters (Section 2.7) can be employed to elute peptides and proteins at different times and hence achieve resolution.

1.4 Aim of chapter

The aim of this chapter is to provide a basic guide to the practice of HPIEC of proteins and polypeptides. The major emphasis will be on small-scale analytical applications. It will be assumed that the reader has a knowledge of or experience in high-performance liquid chromatography. If this is not the case, there are several excellent textbooks on the subject, and these are listed in Appendix I at the end of this chapter. These should be read in conjunction with this chapter, if necessary.

2. MATERIALS AND METHODS

A complete list of items needed to carry out the HPIEC of a set of three standard proteins by anion-exchange chromatography is given as Appendix II at the end of this chapter. A detailed description of the technique for separating these proteins is described in Section 2.15.

2.1 Instruments

2.1.1 Liquid chromatographs

A list of the major manufacturers of high-performance liquid chromatographs is given in the Appendix to this book. Other sources of this information can be found in LC-GC Magazine (3), American Laboratory (4) and in the review by McNair (5).

The most important requirements of a modern liquid chromatograph for HPIEC of proteins and peptides are as follows.

(i) Pump or pumps capable of operating reproducibly between 0.01 and 10 ml/min flow-rates, at up to 6000 p.s.i. pressures.
(ii) A microprocessor based controller to determine the time dependent mixing of two or more buffers.
(iii) An efficient mixer to mix these buffers under the direction of the controller.
(iv) A high-pressure loop injector.
(v) A high-performance ion-exchange column, specifically designed for the unusual requirements of protein chromatography (Section 2.5).
(vi) A flow-through, dual or multi-wavelength detector.
(vii) A recorder/integrator.
(viii) A fraction collector that can preferably be operated at low temperatures (optional).
(ix) An autosampler that can be interfaced with the injector (optional).

Several liquid chromatographs have been designed to be chemically and biologically compatible. These are listed in *Table 1*, along with their major distinctive features. There is little debate concerning the advantages of titanium and glass over stainless steel in their resistance to chemical corrosion. However, there is controversy over the real need for non-ferrous components in avoiding loss of biological activity during HPLC of biopolymers and this is discussed in Chapter 2.

Ion-exchange chromatography of proteins and peptides

Table 1. Liquid chromatographs containing non-ferrous metal wetted components.

Product	Manufacturer	Composition of wetted components	Pressure[a] limit (MPa)	Flow-rates (ml/min)
FPLC^R (Fast Protein Liquid Chromatography)	Pharmacia LKB Biotechnology, Inc.	Glass, fluoro-polymer, polyimide	4	0.017−8.3
GTi HPLC System	Pharmacia LKB Biotechnology, Inc.	Titanium fluoro-polymer, ruby, sapphire	42	0.05−10
Waters 650	Waters Chromatography Division	Fluoro, polymer poly-olefins, ruby, sapphire, quartz	4	0.1−80
IsoPure^{TM} LC System	Perkin Elmer Corporation	Sapphire, polyethylene, titanium, fluoropolymers	42	0.05−10
222Ti	Scientific Systems, Inc.	Ruby, sapphire, titanium, inert polymer	42	0.05−10
Dionex 4000i	Dionex Corporation	Inert polymers	28	0.05−10
Wescan Metal-Free System	Wescan Instruments, Inc.	Inert polymers, sapphire	28	0.01−5

[a]Pressure in Mega Pascals; 10 MPa ≃ 1500 p.s.i.

Selection of a liquid chromatograph can be a very difficult process, since there is a bewildering variety and number of manufacturers and distributors. The following guidelines can be used.

(i) Determine, through requirements of the separation or discussions with experienced practitioners of HPIEC of proteins, the physical limits of the components acquired. For example, maximum pressure; maximum and minimum flow-rates; single or multiple wavelength detection; the need for UV−visible scanning during chromatography; whether a simple flat-bed recorder or computing integrator is sufficient; the need for autoinjector and fraction collector; and the sample volumes required. *Table 2* lists typical operating ranges of pressures, flow-rates, sample volumes, and sample quantities for various size high-performance columns. (See also Sections 2.5−2.14.)

(ii) Refer to manufacturers literature for specifications covering the above features of a liquid chromatograph. The review by McNair (5) is an objective compilation of this information.

(iii) Determine prices, delivery, discounts and very importantly, after-sales service of the instruments.

(iv) Ask experienced protein chromatographers for their advice. These may be in universities, industry, hospitals or the technical support staff of equipment suppliers themselves.

A full description of the methods of installing and operating a liquid chromatograph is, of course, beyond the scope of this chapter. For this, reference to manufacturers instructions combined with discussions with the experienced chromatographer are

Table 2. Commercially available HPIEC column sizes, operating flow-rates, pressures, sample volumes and maximum capacities.

Variable	Commercially available analytical column sizes [i.d. (mm) × length (mm)]							
	4.6 × 30	4.6 × 50	5 × 50	4.6 × 100	4.6 × 250	7.75 × 100	10 × 100	10 × 250
Typical flow-rate (ml/min)	2.0	1.0	1.0	1.0	1.0	1.0	4.0	5.0
Pressure[a] range[c] (10 micron)	NA	NA	0.4–0.9	NA	NA	NA	2.0	NA
Pressure[a] range[c] (7 micron)	NA	NA	NA	NA	NA	NA	NA	NA
Pressure[a] range[c] (5 micron)	NA	1–1.5	NA	2–3	7–10	1–2	3–4	7–10
Maximum recommended sample volumes (ml)	0.3	1–2	1–2	2–5	5–10	5–10	5–10	10–20
Maximum capacities[b] (mg)	0.3	140	100	300	800	800	1200	3000

[a]Pressure in Mega Pascals (10 MPa ≃ 1500 p.s.i.).
[b]Capacities: approximate weight of a protein (of mol. wt 50 000) per ion-exchange column that can saturate all binding sites. Capacities will vary depending on the nature of the bonded phase. Working capacities depend upon resolution of the component of interest, and are typically 20–25% of the maximum capacity.
[c]Pressure ranges measured at typical flow-rate of low ionic strength buffer.
NA = not available.

necessary. However, Section 2.15 contains a brief set of guidelines to follow and important points to observe, in the process of chromatographing proteins.

2.1.2 *pH meters*

pH meters are required in the preparation of buffers and are useful accessories in the monitoring of column effluents. Flow-through cells incorporating a flat-surface pH electrode (i.e. Sensorex, Stanton, CA, USA) are indispensable in the latter technique. The electrical output from such a device can be conveniently monitored with a standard flat-bed chart recorder. Simply attach the electrode to a meter (e.g. Model 701A from Orion, Cambridge, MA, USA). Then connect wires from the meter to the recorder where the pH will be monitored simultaneously at the meter and output as a trace of pH as a function of time.

2.1.3 *Conductivity meters*

These give a qualitative measure of ion strength of the ion-exchange buffers, either before or after the HPIEC column. The constant conductivity of the column effluent is the sign that ionic strength equilibration has been reached. Complete equilibration, however, is reached only when the ionic strength and pH are constant. Conductivity cells and meters suitable for flowing systems are available from companies such as Amber Science, San Diego, CA, USA.

2.1.4 *Spectrophotometers*

Standard spectrophotometers are used to measure absorbance of protein solutions when recoveries are being determined (Section 2.17). Recording spectrophotometers are sometimes useful in measuring the spectra of purified proteins for identification purposes. The spectral properties of the buffer solutions are rarely measured, although the variation of absorbance with the mixing of two buffers at a single wavelength (baseline) is often performed using the UV detector of the liquid chromatograph.

2.1.5 *Centrifuges*

Small volumes of solutions (1 or 2 ml) can be clarified by rapid (several minutes) centrifugation with an Eppendorf centrifuge. Large sample volumes require a high speed centrifuge. Centrifugation is one method of clarifying a sample before injecting it onto a high-performance column (Section 2.1.6).

2.1.6 *Filtering aids*

Careful filtration of buffer solutions will enhance the life-time of the pumps and HPIEC columns. A wide range of devices for large and small volume filtration (preferably through 0.45 micron filters) are available from companies such as Millipore and Gelman.

Filtration of samples is one clarification method critical to the life-time of a column. Particulate material may bind to the inlet frit and the top of the column causing high back-pressure and occasionally changes in the chromatography itself.

The placement of a 0.5 micron high-pressure filter between the injector and column offers important protection of the latter. It is important to monitor the system pressure

Table 3. Common buffers and salts used in HPIEC.

Buffer	pH range	Buffer	pH range	Neutral salts
H_3PO_4	1-3	Na citrate	3-5	NaCl
KH_2PO_4	6-8	Na glycinate	8-10	Na_2SO_4
NH_4 acetate	6-8	Acetic acid	4-5	$(NH_4)_2SO_4$
Na acetate	4-5	Morpholine		
Tris-HCl	7-9	Ethane sulphonic acid (MES)	5-7	

to determine when it is necessary to clean or replace such filters. As a general guide this should be carried out when the column back-pressure has increased by 50% over its initial value.

Dialysis is an ultrafiltration process whereby ions or molecules can be selectively removed from a solution. Flexible tubing for this technique can be obtained which will allow ions below a certain molecular weight to pass through. The technique is useful in reducing ionic strength (de-salting), buffer exchange and removing low molecular weight impurities from a sample but it is a slow process.

2.1.7 *Waters for buffers and solutions*

For most buffers a source of sufficient volumes of de-ionized water is necessary. Further purification of this water will generally only be required in order to eliminate endotoxins (pyrogens). In-lab purification systems are necessary in this case in order to control potential sources of contamination more effectively.

2.2 **Buffers, salts and mobile phase additives**

The resolution of a protein or proteins of interest from a sample depends, among other things, upon the specific buffer ions, their ionic strengths and the pH of their solutions. Buffers are required to control pH as much as possible, and salts act to displace the biopolymer from the surface of the ion-exchange material. Buffers and salts that are commonly used in HPIEC are listed in *Table 3*.

Other mobile phase additives are used in HPIEC in order to maintain certain properties of proteins in solution. *Table 4* gives a list of these reagents and their general purpose.

In order to take full advantage of the changes in selectivity that can be achieved by using different buffers, salts and pH ranges, it is recommended that stocks be kept of all the common reagents listed in *Table 3*. If possible, obtain the highest quality chemicals that can be afforded, especially if working at 210-220 nm wavelength detection, where many impurities absorb strongly.

Adjust pH values using the conjugate acid or base to the major buffer ion. For example, use phosphoric acid to reduce the pH of a KH_2PO_4 solution.

All mobile phases should be filtered through 0.45 micron membranes, then de-gassed before use, as in other forms of high-performance liquid chromatography.

Buffers and salt solutions should be stored at 4°C for 2-3 days at most when not in use, then made up freshly, in order to avoid contamination with microorganisms. In demanding purifications, endotoxin-free water may be used. Some buffers are available which are guaranteed to be free of several specific classes of enzymes, including

Table 4. Mobile phase additives for HPIEC.

Reagent	Purpose
Urea	Denaturant, solubilizer
Guanidine hydrochloride	Denaturant, solubilizer
Sodium dodecyl sulphate	Denaturant solubilizer
Glucose, sucrose, polyethylene glycol	Structure stabilizer
Dithioerythreitol	Reducing agent
EDTA	Sequesters calcium, magnesium
Mercaptoethanol	Reducing agent
Neutral detergents	Solubilizers

endonucleases, ribonucleases and proteases. Vessels containing buffer solutions should be sealed against airborne contaminants during use.

Buffers should be made up according to the following procedures.

(i)   Calculate the weight of a given salt needed to make up the required volume of buffer of the required molarity. Weigh this into a flat bottom beaker.

(ii) Dissolve this quantity into approximately half the final volume of de-ionized water, with magnetic stirring.

(iii) Make up to 95% of the final volume with de-ionized water. Adjust the pH of this solution.

(iv) Make up to the final volume and re-check the pH of this solution.

(v)   Filter and de-gas this solution.

(vi) Pour this solution into containers which can be capped or sealed with Parafilm.

(vii) Check carefully each day for the appearance of cloudiness or the growth of microorganisms in the buffers. Such solutions must be discarded.

2.3 **Protein standards**

2.3.1 *Introduction*

Trial separations carried out with mixtures of protein standards are a practical way of getting started in HPIEC. Samples must be prepared, injected, collected and stored using techniques similar to those for a real-life sample. The added advantage with standards is that greater control exists over the composition and content of the sample. Furthermore, measurements of capacity, mass and activity recoveries are simpler and more accurate with known standard mixtures. In addition, the optimization of chromatographic parameters (Section 2.16) to maximize resolution, recovery and throughput, is best carried out initially using standard proteins of interest.

2.3.2 *Suggested proteins for HPIEC*

Table 5 lists the names and isoelectric points of proteins suitable for learning the basic techniques of HPIEC. Several sources of these materials are listed in Appendix III at the end of this chapter.

Table 5. Standard proteins for HPIEC.

Name	Mol. wt	pI
Insulin	5700	5.3
Cytochrome *c*	13 000	10.6
Ribonuclease	14 000	7.8
Hen egg white lysozyme	14 300	11.0
Horse myoglobin	17 000	7.0
Human growth hormone	21 500	6.9
Soybean trypsin inhibitor	22 460	NA
Carbonic anhydrase	30 000	7.3
Carboxypeptidase	34 000	6.0
Pepsin	35 500	<1.0
Ovalbumin	40 000	4.6
Haemoglobin, horse	65 000	6.9
Human serum albumin	66 500	4.8
Bovine serum albumin	67 000	NA
Lactic acid dehydrogenase	140 000	NA
Immunoglobulins G	150 000	$6.4-7.2$
Catalase	250 000	5.6
Fibrinogen	330 000	5.5
Urease	480 000	5.1
Thyroglobulin	660 000	4.6

NA = not available.

2.3.3 *Preparation of standard solutions*

Most proteins are obtained as a lyophilized (dried) powder consisting of up to 100% protein. Specifications will indicate what proportion is the protein of interest, and how much buffer salt and other protein is present. Other proteins may only be available as a suspension in salt. Store opened and unopened bottles in a refrigerator, preferably in alphabetically labelled small boxes or drawers.

The general procedure for making up individual solutions is as follows.

(i)   Proteins should preferably be dissolved in a low ionic strength buffer such as 25 mM KH_2PO_4, at pH 7. The pH and/or ionic strength may need to be altered to achieve dissolution.

(ii)   Weigh a 25 ml volumetric flask. Weigh $50-100$ mg of a standard protein into the flask.

(iii)   Add 5 ml of buffer at room temperature and shake gently. Add more, if necessary, to dissolve the protein, and make up to the mark.

(iv)   Freeze the solution till required. Do not re-freeze once the solution has thawed. Once the solubility characteristics of a standard are known, prepare concentrated solutions and dilute to working strength. Freeze multiple samples ($1-5$ ml), but do not re-freeze once they have thawed.

(v)   Seal each tube with Parafilm® .

(vi)   Label each container of solution carefully with an indelible marker or a small piece of coloured, adhesive-backed tape.

(vii)   Spin down any sediment or undissolved portion using a small centrifuge, before freezing or before use.

Table 6. Mixtures suitable for achieving base-line resolution.

Anion-exchangers	Cation-exchangers
Cytochrome *c* (unretained)	Ovalbumin (unretained)
Conalbumin	Soybean trypsin inhibitor
β-Lactoglobulin B	Haemoglobin
β-Lactoglobulin A	Cytochrome *c*
Calmodulin	Lysozyme

Table 7. Biological sources of proteins.

Sources	Major proteins
Hen egg white	Ovalbumin, lysozyme
Milk	β-Lactoglobulin A, B; α-lactalbumin
Human plasma	Human serum albumin; α, β and δ globulins; fibrinogen

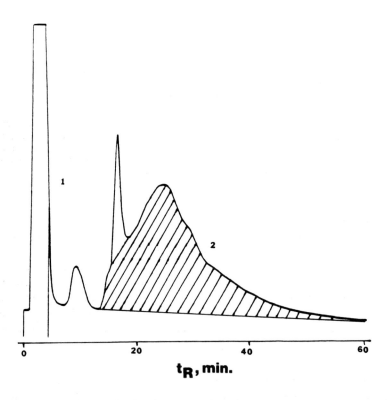

Figure 1. Separation of immunoglobulins from human plasma using a mixed-mode ion-exchanger. Column, 7.75 mm × 100 mm, 5 micron BAKERBOND ABx; mobile phase, initial buffer (A) 25 mM MES, pH 5.4; final buffer (B) 500 mM $(NH_4)_2SO_4$ plus 20 mM sodium acetate, pH 6.3; gradient, 0% B to 25% B over 60 min; flow-rate, 1.0 ml/min; detection: UV (280 nm), 1.0 AUFS; sample, 2 ml (0.5 ml crude human plasma diluted to 2 ml with buffer A. Peaks: **1**, human serum albumin; **2**, immunoglobulin fraction (shaded).

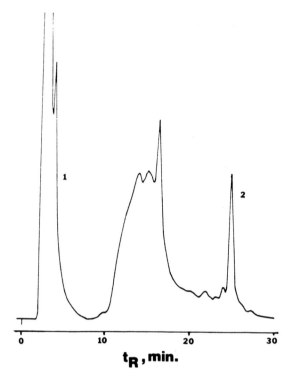

Figure 2. Purification of lysozyme from crude hen egg white using a weak cation-exchanger. Column, 4.6 mm × 250 mm, 5 micro BAKERBOND WP*-CBX*; mobile phase, initial buffer (A) 10 mM MES, pH 5.6; final buffer (B) 1.0 M sodium acetate, pH 7.0; gradient, 0–100% B over 30 min; flow-rate, 1.0 ml/min; detection: UV (280 nm), 1.0 AUFS; sample, 0.05 ml [fresh egg white, 30 mg dissolved in buffer A (100 ml) and carefully filtered]. Peaks: **1**, ovalbumin; **2**, lysozyme.

2.3.4 *Multiple component mixtures*

Mixtures which are usually suitable for achieving base-line or near base-line resolution, on good quality HPIEC columns, are listed in *Table 6*.

Each standard solution should be chromatographed separately under standard conditions (Section 2.15). From the peak heights, an estimate can be made of the proportions of individual solutions that will be required to make up the mixture. Examine the mixture for precipitates and centrifuge if necessary. All single protein solutions and mixtures should be kept at about 4°C at all times. Chromatography can be carried out at room temperature (10–30°C).

2.4 **Real-life samples and their preparation**

Sample handling and preparation techniques associated with HPIEC of proteins and peptides, are critical to the success of the separation. The field of sample preparation for biopolymer separation has been reviewed by Wehr (6) and Majors (7). Further details can be found in works by Suelter (8), Brewer, Pesce and Ashworth (9); and in the subject indexes of *Methods in Enzymology* (10).

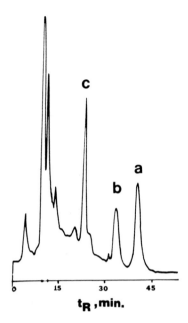

Figure 3. Separation of milk proteins using a weak anion-exchanger. Column, 7.75 mm × 100 mm, 15 micron BAKERBOND WP-PEI; mobile phase, initial buffer (A) 10 mM Tris, pH 7.0; final buffer (B) 2 M sodium acetate, pH 6.0; gradient, 0–100% B over 60 min; flow-rate, 1 ml/min; detection: UV (280 nm), 0.5 AUFS; sample, 0.05 ml (LIPOCLEAN®-extracted skim milk, diluted three times with buffer A). Peaks: **c**, conalbumin; **b**, β-lactoglobulin B; **a**, β-lactoglobulin A.

Relatively simple biological sources of proteins may be experimented with as the next step in the technique of HPIEC. *Table 7* gives a suggested list of these samples. *Figures 1, 2, 3* and *4* illustrate the characteristic chromatography of these materials, and the chromatographic conditions used.

The method of preparation of these samples is as follows.

(i)   Dilute each sample with the buffer that will be used to bind the components to the HPIEC column (buffer A in the legend to *Figures 1, 2* and *3*).

(ii)   Egg white should be carefully mixed with a broad, flat stirrer, and the resulting solution centrifuged to remove the flocculant material.

(iii)   Extract the diluted milk solution and the human plasma with an equal volume of LIPOCLEAN™ (11) to remove most lipids.

(iv)   Centrifuge the solutions of milk and plasma. Large volumes can be carefully filtered.

2.5 Column selection

A general approach summarizing the steps to be used in choosing a column for the ion-exchange HPLC of peptides and proteins is given in *Table 8*. These are described in detail in the following sections.

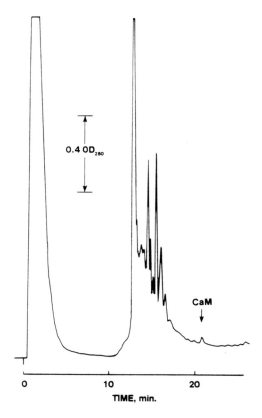

Figure 4. Separation of calmodulin from bovine brain extract using a weak anion-exchanger. Column, 4.6 mm × 50 mm, 5 micron BAKERBOND WP-PEI; mobile phase, initial buffer (A) 25 mM Tris, pH 7.0; final buffer (B) 2.0 M sodium acetate, pH 5.0; gradient, 0−100% B in 15 min; flow-rate, 1 ml/min; detection: UV (280 nm); sample, 1.0 ml (1.0 g bovine brain extract dissolved in 100 ml of buffer A and centrifuged. Protein concentration ~8 mg/ml.) Peaks: CaM, calmodulin.

Table 8. Five major steps in column selection.

1. Characterize sample [matrix + component(s) of interest]
2. Characterize component(s) of interest
3. Choose type of ion-exchanger/pH range
4. Choose column configuration
5. Examine quality, cost, service

2.6 **Sample characterization**

The properties of the sample that are most important in column selection for ion-exchange HPLC are as follows: pH, ionic strength, pyrogen content, indicator dyes, volume, matrix constituents (lipids, proteins, nucleic acids, for example).

Table 9. Important properties of proteins and peptides.

Property of component(s) of interest	Utility
pI (isoelectric point)	Determines choice of anion- or cation-exchanger
Molecular weight	Determines desirable pore size
Hydrophobicity	Extent of possible hydrophobic binding to ion-exchanger
Reactivity (oxidation/reduction, catalysts)	Inhibitors may be required
Degree of aggregation	Oligomers may need to be removed; further aggregation may need to be prevented, disaggregation may be required
Concentration	May determine ease and scale of purification
Solubility	Must be soluble in ion-exchanger buffers
Stability	May determine temperature and speed of operation necessary
Microheterogeneity	May require highly resolving techniques

The pH and ionic strength will normally need to be adjusted before ion-exchange chromatography, principally in order to achieve binding. Unwanted pyrogens may be removed by dialysis after chromatography or by the choice of a bonded phase that binds the component(s) of interest but does not bind the pyrogen. Dyes (such as phenol red) that indicate pH may be a component of biological tissue culture samples. These must be removed during or after chromatography. The sample volume, when taken in conjunction with the total mass of bindable material, the capacity of the column (see Section 2.8) and whether purification or analysis is intended, will determine the size of the column. Many ion-exchangers exhibit sufficient hydrophobic character to bind lipids even at low ionic strength. Immiscible solvents such as LIPOCLEAN^{TM} may be used to extract non-polar constituents before chromatography and without harming the activity or solubility of the peptide or protein of interest.

2.7 **Component characterization** (see also *Table 5*)

The properties of the peptide or protein of interest that are most important in column selection are as follows: pI, molecular weight, hydrophobicity, reactivity.

The importance of these properties is summarized in *Table 9* and amplified in Section 2.8.

2.8 **Choice of ion-exchanger**

The properties of the ion-exchange medium that must be considered are as follows.

(i)   Charge of surface (especially charge as a function of pH).
(ii) Nature of substrate (particle size, pore size, chemical nature, surface area, resistance to deformation).
(iii) Nature of bonded phase (resolving power, scale-up capability, stability, ease of cleaning, regeneration and sterilization).

pK$_a$ ⧙⧙ pH ⧙⧙ pI

pK$_a$ of fixed charge group
pI of protein/peptide

Figure 5. A schematic guide to the use of cation- and anion-exchangers. The operational pH of the mobile phase must be between the isoelectric point (pI) of the protein or peptide and the pK$_a$ of the ion-exchanger acid groups. The acidic groups may be -SO$_3$H (sulpho), -COOH (carboxy), -OP$_3$H$_2$ (phospho), -NH$^+$ (primary, secondary, tertiary ammonium). In anion-exchange, pI < pK$_a$, and a cation-exchange, pI > p K$_a$.

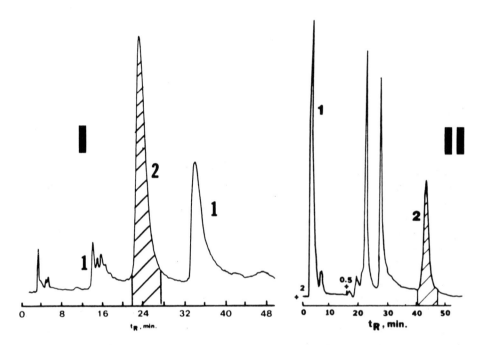

Figure 6. Separation of a monoclonal antibody from mouse ascites fluid using weak anion-exchange (**I**) and mixed-mode exchange (**II**) chromatography. Conditions for chromatogram I: column, 4.6 mm × 250 mm, 5 micron BAKERBOND ABx; mobile phase, initial buffer (A) 10 mM potassium dihydrogen phosphate, pH 6.80; final buffer (B) 500 mM potassium dihydrogen phosphate, pH 6.4; gradient, 0−25% B over 60 min; flow-rate, 1 ml/min; detection: UV (280 nm); 0.2 AUFS; sample, 0.2 ml (mouse ascites fluid, 40 μl, diluted with 160 μl of buffer A). Conditions for chromatogram II: column, 4.6 mm × 250 mm, 5 micron BAKERBOND ABx; mobile phase, initial buffer (A) 10 mM potassium dihydrogen phosphate, pH 6.0; final buffer (B) 250 mM potassium dihydrogen phosphate, pH 6.8; gradient, 0−50% B over 60 min; flow-rate, 1 ml/min; detection, UV (280 nm) 2.0 AUFS (flow-through) and 0.5 AUBS (bound peaks); sample, 0.5 ml (mouse ascites fluid, 100 μl, diluted with 400 μl of buffer A). Peaks: **1**, albumins, transferrin; **2**, monoclonal immunoglobulins (IgG class).

2.8.1 Charge of surface

When the range of pH values is known, over which the peptide or protein of interest is stable, the bonded phase surface charge may be chosen so as to bind the component via electrostatic forces only (if possible). The isoelectric point (pI) of the constituent should be known, and the type of ion-exchanger and operational pH range selected according to the guide in *Figure 5*.

Table 10. Grouping of ion-exchangers based on functional group type.

Anion-exchangers	Typical functional group	Cation-exchangers	Typical functional group	Mixed bed exchangers
Weak	Polyamino	Weak	Carboxymethyl	Amino + carboxy
Moderate	Diethylaminoethyl	Moderate	Phospho	Carboxy + sulpho
Strong	Quaternary ammonium	Strong	Sulpho	Quat + sulpho

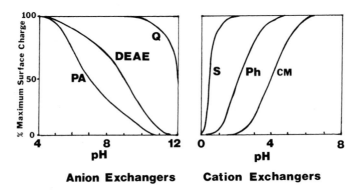

Figure 7. Idealized representation of the variation of surface charge with pH for various acid and basic groups on ion-exchangers. PA, polyamine, DEAE, diethylaminoethyl; Q, quaternary ammonium; CM, carboxymethyl; Ph, phospho; S, sulpho.

At pH values greater than pI, a peptide or protein is generally negatively charged and at pH values less than the pI, these materials are usually positively charged. Thus a given peptide or protein may be chromatographed on both an anion-exchanger or a cation-exchanger, depending upon the pH of the sample and mobile phase and whether the surface is appropriately charged at this pH. In general, however, acidic proteins (pI < 6) are chromatographed when they are negatively charged; basic proteins (pI > 8) are chromatographed when they are positive; and proteins whose pI lies between 6 and 8 can be chromatographed as either negative or positive species on anion-exchangers or cation-exchangers, respectively (*Figure 6*).

It is important to know or appreciate the manner in which the surface charge of an ion-exchanger varies as a function of pH and ionic strength. The last parameters are those most commonly chosen to control efficient binding and elution. For weak and moderate ion-exchangers the lower the pH the more positive the surface becomes, and the higher the pH, the more negative or neutral the surface becomes. For true strong ion-exchangers, however, surface charge is independent of pH within the range 2 − 12. A knowledge of the magnitude of the surface charge of the ion-exchanger and that of the polypeptide or protein at the same pH and ionic strength, permits a rational choice of an ion-exchanger to bind a given biopolymer. In general conditions must be chosen to optimize the strength of binding between oppositely charged protein and ion-exchanger. The aim in resolving a component biopolymer of interest from a complex mixture, is to control its binding strength so that it is different to those of all other components.

106

Table 11. Physico/chemical requirements for high-performance ion-exchange substrates.

Property	Necessary	Desirable
Structural integrity (resistance to deformation)	Yes	Yes
Microparticulate nature	Yes	Yes
Porosity	No	Yes
Range of pore sizes	No	Yes
Lack of solubility in mobile phases	No	Yes
Surface hydrophilicity	Yes	Yes
Chemical reactivity	Yes	Yes
Chemical resistance	Yes	Yes
Microbial resistance	No	Yes

The functional groups that determine the variation of surface charge with pH, can be used to define the various classes of ion-exchangers, as given in *Table 10.*

The actual shape and position of the plots of the variation of surface charge with pH shown in *Figure 7* will depend upon the total environment in which the functional groups are located, which in turn will usually depend upon the manner of synthesis, the substrate and the manufacturer.

The use of the terms 'strong', 'moderate' and 'weak' still appears to be a subject of significant confusion and controversy, when applied to ion-exchange chromatography. The origin of these terms lies clearly with the definitions of strong, weak and moderate Brønsted acids and bases corresponding to the functional groups that are bound to the ion-exchange surface. The average pK_a and pK_b values of the various acid and basic groups set the boundaries beyond which there is more charge or less charge. The strength of attraction between the fixed charges on the ion-exchanger and mobile ions being exchanged is not implied in these terms.

2.8.2 *Nature of substrate*

It is only relatively recently that the technology has been developed to produce particulate materials that meet the requirements of HPIEC of peptides and proteins. *Table 11* lists these requirements. Historically the first of these substrates was silica, followed by polystyrene highly cross-linked with divinylbenzene, then hydroxylated polyether, polymethacrylates, cross-linked with glycerol and glycols, alumina and most recently, a polyester−polyamine co-polymer.

The major requirement that spurred the development of the above materials for HPLC was that of structural integrity. Columns packed with small particles, that are necessary for high efficiencies, must generally be able to withstand high pressures with little or no deformation of the particle's structure. High pressures (several thousand p.s.i.) are required during the slurry packing of such columns. Furthermore, rapid re-equilibration of HPLC columns is best accomplished at high linear velocities and consequently high pressures.

The requirement for small diameter particles in high-resolution chromatography is well established. The great majority of HPLC columns for analysis contain particles which are 10 micron or less in diameter. HPLC columns used in preparative work contain particles that are rarely greater than 25 micron, which may be considered the upper

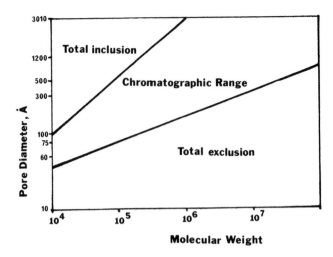

Figure 8. Pore diameter limits for total exclusion and total inclusion of polymers as a function of their molecular weights.

limit of the diameter of so-called 'high-performance' packings. Beyond this figure it is generally agreed that the term 'medium-performance' should be used. It is also at about 25 micron that columns containing such packings may be efficiently dry-packed.

The great majority of HPLC bonded phases are prepared on porous particles. The major advantage of such substrates is their high specific surface area. Binding capacity of a given peptide or protein is dependent upon the proportion of this surface that can be accessed by the biopolymer. Ion-exchangers prepared from porous substrates can bind up to 40% of their own weight of an average sized (mol. wt 40 000) protein. Furthermore, such a high capacity means that complex samples containing low concentrations of a protein of interest may be safely analysed and purified with a low risk that this protein will not bind. Several bonded phases suitable for the ion-exchange HPLC of peptides and proteins, have been developed using non-porous substrates. The main advantage of these is the speed with which separations occur. This is due to the fact that the large polymers do not need to diffuse into and out of pores. The main limitation of such supports is their very low capacity, which results in rapid overloading and peak-broadening unless due care is taken. Thus porosity is not necessary, but is desirable.

There is general agreement that pore diameter must be greater than a given size to be suitable for chromatography of a given protein. *Figure 8* indicates the pore size limits of bonded phases as a function of molecular weight.

Low solubility of the ion-exchanger substrate in mobile phases and cleaning/regenerating/sterilizing solutions is, of course, desirable (Section 2.8.1).

Hydrophilic (readily hydrated) surfaces are generally considered necessary for ion-exchange, since such environments are energetically favourable for the formation and breaking of only ionic bonds. The presence of non-polar regions in an ion-exchanger may promote dispersive and dipole–dipole attractive forces with non-polar regions on the peptide or protein. The simultaneous operation of these several disparate forces

Table 12. General nature of major substrates used in HPIEC of proteins and peptides.

Name	Structure	P max (mPa)	pH range	Usable organic solvents
Silica	$O-\overset{\overset{O}{\|}}{\underset{\underset{O}{\|}}{Si}}-O$	100	2 – 8	All
Polystyrene – divinylbenzene	$-C-\overset{\overset{O}{\|\|}}{C}-C-$ $(CH_2)_2-\langle\bigcirc\rangle-(CH_2)_2$ $C-\overset{\overset{\|}{}}{C}-\overset{\|}{C}$ $\overset{\|\|}{O}$	40	2 – 12	Most
Hydroxylated polyether	$O-CH_2-\overset{\overset{OH}{\|}}{CH}-CH_2-O-$	10	2 – 12	Most
Polymethyl methacrylates	$(-CH_2-\overset{\overset{COOCH_3}{\|}}{\underset{\underset{CH_3}{\|}}{C}}--)_n$	10	2 – 12	Most
Alumina	$\overset{O\diagdown \quad \diagup O}{\underset{\underset{O}{\|}}{Al}}$	100	4 – 14	All
Polyamine polyester	$(-CH_2-CH_2-NH-)_n$ (ester structure not available)	40	2 – 12	Most

on the binding of a polymeric ion, has undesirable consequences for ion-exchange chromatography. Band width generally increases and mass recovery may decrease.

The substrate must be chemically reactive in order to produce the fixed charged groups that are bonded covalently or electrostatically to the surface. However, very little detailed chemistry has been published in the patent or other literature concerning the reactivity of substrates for commercially produced high-performance ion-exchangers.

Resistance of the substrate to chemical reaction (inertness) is important in determining the lifetime, re-usability, regenerability, cleaning and sterilizability of the ion-exchange material. Naturally the chemical inertness required is that existing under normal chromatographic conditions:

(i)   pH values from 2 to 10;
(ii)   ionic strengths from 0 to up to 3 M; and
(iii)   possible presence of organic solvents.

The substrate cannot usually be entirely inert, since some reactivity must be present in order to introduce charged groups such as sulpho and quaternary amino.

Table 12 gives a list of the major high-performance substrates, together with general conditions under which the matrix is stable.

Microbial and enzymatic resistance is desirable for a number of reasons.

(i)   Most biological fluids contain enzymes which catalyse various reactions of reactive substrates such as polysaccharides. These processes not only tend to degrade the structural integrity of the substrate, but where microorganisms are involved, may cause the release of endotoxins into the product stream.

(ii) Accumulation of microorganisms within a packed column often causes increased back-pressure, preventing further separation and purification.

(iii) Concerns about the above reduce column life-time and increase maintenance of high-performance columns.

Currently, all but one (Superose^TM) commercial high-performance substrates are synthetic materials, exhibiting minimal sensitivity to microbial or enzymatic attack.

2.8.3 *Nature of bonded phase*

The properties of a high-performance ion-exchanger are comprised of the combined properties of the substrate and surface. Important factors in HPIEC that relate to the nature of the bonded phase as a whole are discussed below.

(i) *Resolving power and efficiency.* The resolving power of a given HPIEC column is a difficult parameter to measure. The extent of resolution of a given pair of proteins (for example the A and B forms of β-lactoglobulin) will depend upon the mobile phase ion species, pH, the surface charge density on the bonded phase, the column efficiency for large molecules and the gradient steepness. Efficiency is determined primarily by particle diameter (see Section 2.8.2). The retention time between the two peaks depends upon the difference in ionic strengths required to displace each protein. This is turn depends upon the mode of interaction of the binding region on the protein with the surface of the ion-exchanger (see Introduction). A physically flexible surface will allow this attractive interaction to be optimal for each protein. Consequently such an ion-exchanger will be most sensitive to small differences in the surface topography of otherwise similar proteins and be best able to resolve these biopolymers.

Thus, in selecting a column for HPIEC, a soft surface, rigid substrate packing with a small particle diameter, should be optimum.

(ii) *Scale-up capability.* Multi-milligram quantities of peptides or proteins are readily purified using analytical-sized columns (4.6 × 250 mm) and packings of 5 or 10 micron particle diameter. To purify larger quantities, larger columns may be required, which may contain larger particles. It should be determined whether such columns will be required for future work and, if so, whether they are available from the same manufacturer. Some commercial sources of high-performance ion-exchange media offer a family of products suitable for a full range of applications from nanogram to kilogram quantities of peptide or protein.

(iii) *Stability.* Ideally, maximum bonded phase stability would be achieved where the charged groups are as integrated as possible with the entire structure of the packing material, and where no groups exist that can react with water, oxidants, reductants or enzymes capable of catalysing such reactions.

It is well established that synthetic polymers used in HPIEC, such as derivatized polystyrenes, polyhydroxyethers and polymethacrylates, are generally stable to hydrolysis over the pH range 2−12. In addition, Chicz and co-workers (12) have demonstrated that ion-exchangers made from high-performance alumina and zirconia-coated silica, have negligible solubilities at pH 14. Zirconium-treated silica patented by Dupont is known to withstand pH values up to 8.5 with little increase in solubility. The BAKERBOND* series of a polymer-coated silica-based, ion-exchangers and hydrophobic interactors, has general stability over the pH range 2−10.

Stability of an ion-exchanger at high pH may be important when alkaline clean-up washes are used to remove final traces of polypeptides or proteins, or the lipopoly-saccharides that are the major class of endotoxins. However, alternate, less harsh methods have been described to achieve this clean-up (see below).

(iv) *Re-equilibration*. At the end of an elution in which pH and ionic strength may be different from the starting conditions, re-equilibration will be necessary. A high-performance ion-exchanger can be said to be re-equilibrated when the ionic strength and pH of the solution within and without the pores of the packing (if porous) are the same as those at the beginning of the separation. Furthermore the state of ionization of the charged surface groups must be the same as that reached initially.

Re-equilibration will be faster for strong ion-exchangers since there will be no change in the state of ionization of the charge surface groups. In general, the process of re-equilibration is best carried out at a higher flow-rate than that of the chromatography since time will be saved. Naturally the pressure and flow-rate limits of the column must not be exceeded.

(v) *Clean-up*. Modern packings for HPIEC should show mass recoveries of greater than 90% for a wide range of peptide and protein classes (see Section 2.17 for a discussion of recoveries). Even in those cases where recoveries approach 100%, there will usually be small quantities of biopolymer that need to be removed either after each separation or after a given number of cycles. The efficiency with which a clean-up

Table 13. Clean-up reagents for BAKERBOND WIDE-PORE* columns for HPIEC.

Reagents (in order of priority)
1. Sodium acetate (2 M, pH 8)
2. Acetic acid (1%)
3. DMSO/water (50/50)
4. Chaotropic agents such as aqueous urea, guanidine hydrochloride (5−6 M) if necessary

Protocol for use of above reagents
1. Wash the column with 10 column vols of water.
2. Run a gradient with 2 M sodium acetate, pH 8 (10 vols).
3. Follow this with water (5 vols); then acetic acid (1%, 10 vols).
4. Repeat steps 1−3 if a single clean-up is insufficient.

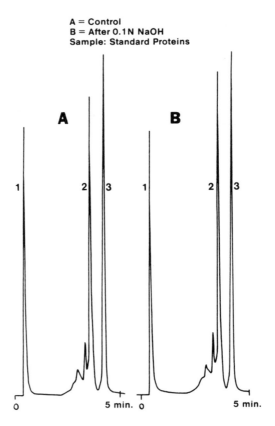

Figure 9. Separation of standard proteins using weak anion-exchanger, before and after treatment with 36 column vols of 0.1 NaOH. Column, 4.6 mm × 50 mm, 5 micron BAKERBOND WP-PEI; mobile phase, initial buffer (A) 10 mM potassium dihydrogen phosphate, pH 6.8; final buffer (B) 500 mM potassium dihydrogen phosphate, pH 6.8; gradient, 0–100% B over 2 min; flow-rate, 1 ml/min; detection, UV (280 nm) 0.1 AUFS. Peaks: **1**, cytochrome *c*; **2**, ovalbumin; **3**, β-lactoglobulin A. Chromatogram **A**: control chromatography. Chromatogram **B**: same chromatography after 30 ml (36 column vols) of sodium hydroxide (0.1 M) was cycled through column, neutralized with buffer B to pH 6.8 and re-equilibrated.

step may be performed is an important consideration in the choice of a bonded phase. Several commercial suppliers recommend clean-up procedures as part of the use and care of their products. *Table 13* lists the reagents and brief procedures that can be used for clean-up of the BAKERBOND WIDE-PORE* family of bonded phases. *Figure 9* illustrates the chromatographic elution profile obtained from a real-life sample before and after clean-up with sodium hydroxide. The use and care of HPLC columns for biopolymer analysis has been reviewed by Wehr (13).

Sterilization is the process by which living cells are killed or prevented from reproducing. HPIEC columns that are compatible with ethanol, methanol, acetonitrile, 0.1% sodium azide solution or 70/30 (v/v) ethanol/water, may be washed with 10 column volumes of these solutions to effect sterilization. The manufacturers' instructions should be followed in all cases.

Table 14. Major commercial column dimensions, advantages, limitations, application areas.

Dimensions (i.d. × length mm)	Advantages	Limitations	Application areas
4.6 × 30	Low pressure/high speed	Low capacity/decreased resolution	Fast analysis
4.6 × 50	Low pressure/high speed	Low capacity/decreased resolution	Fast analysis
5.0 × 50	Low pressure/high speed	Low capacity/decreased resolution	Fast analysis
4.6 × 100	Moderate pressure and speed	Decreased capacity	Moderate speed analysis
4.6 × 150	Moderate pressure and speed	Fair capacity	Moderate speed analysis
4.6 × 250	High resolution and capacity	High pressure	High resolution analysis, semi-preparative
7.75 × 100	Low pressure/high capacity	Decreased resolution	Fast analysis, semi-preparative
10.0 × 100	Low pressure/high capacity	Decreased resolution	Analysis, semi-preparative

2.9 Column configuration and construction

The choice of column configuration is dictated by the manufacturer of column hardware or by commercially available pre-packed columns. The first columns for analytical HPIEC were 25−30 cm long and 4−5 mm i.d. More recently, shorter columns (3.3 cm long) have become commercially available. *Table 14* lists the major column sizes available with a description of their advantages, limitations and important areas of application.

The construction of an HPIEC column is important when considering chemical inertness, biological compatibility, adaptability to various chromatographs, pressure tolerances and, occasionally, ease of disassembly for column repair when necessary. In addition, column hardware design will influence its chromatographic efficiency somewhat, although this latter parameter depends far more upon the nature of the packing medium and the sophistication of the column packing technique. *Table 15* lists the major materials of construction of HPIEC columns together with brief comments on their physical, chemical and biological compatibility with mobile phase components, instrumentation and peptides and proteins of interest.

Tables 14 and *15* can be used to select the most appropriate configuration and construction of a column. Naturally the choice will be limited to the column types offered by the manufacturer, although some companies offer a custom column service.

2.10 Quality, cost and service

These aspects of bonded phase selection are, of course, very difficult to assess. However, a brief, necessarily incomplete general guide to this topic can be given.

Table 15. Major column construction materials.

Material	Physical strength	Chemical compatibility	Biological compatibility
End-fitting			
316 Stainless steel	Highest	Avoid halides	Compatible
Titanium	High	Non-corrodible	Compatible
Fluoropolymer	High	Non-corrodible	Compatible
Frits			
316 Stainless steel	High	Avoid halides	Compatible. Adsorbs protein slightly
Titanium	Moderate	Non-corrodible	Compatible. Adsorbs protein slightly
Column wall			
316 Stainless steel	Highest	Avoid halides	Compatible. Adsorbs protein slightly
Borosilicate glass	Low	Non-corrodible	Compatible. Adsorbs protein slightly
Fluoropolymer	High	Non-corrodible	Compatible. Slight adsorption of protein

2.10.1 *Quality*

Bonded phases should be reproducible, sufficiently resolving, stable, easy to clean, regenerate and sterilize and give high mass and activity recoveries. Some or all of this information may be obtainable from the manufacturer. Each manufacturer's weak anion-exchanger, for example, will be different from the others. This may be useful because two components, poorly resolved on one ion-exchanger, may successfully separate on another.

Objective assessments of quality can be obtained from colleagues or independent workers who have published their work. Their advice is very important in deciding on quality.

2.10.2 *Cost*

Take into account the specific binding capacity since several available HPIEC columns can be used for both analysis and small prep-scale work. Cost should, of course, be related to quality (Section 2.10.1). Catalogues and price lists from all major suppliers should be obtained and kept conveniently together.

2.10.3 *Service*

Ascertain the depth of technical knowledge available in a company to support their bonded phase products. This knowledge is usually found in their customer service department, technical sales or marketing specialists, and research and development scientists. The name of the marketing people or scientists of most value in this respect, will be found on papers published on the products in which you are interested. These people will be able to give detailed information on column use and care, applications, and in most cases, can help in solving new separation problems and troubleshooting

Table 16. Commercially available wide-pore HPIEC columns for anion-exchange of proteins and peptides.

Packing	Manufacturer	Strong	Weak	Pore size (nm)	Particle diameter (μm)
AQUAPORE®	Brownlee	No	Yes	30	7
BAKERBOND WIDE-PORE*	J.T.Baker	Yes	Yes	30	5
Chemcosorb	Chemco	Yes	No	30	7
Daltosil, Si	Serva	No	Yes	30, 10	5, 10
Dynamax® -300 A AX	Rainin	No	Yes	30	12
Hydrophase™	Interaction Chem	No	Yes	∼100	10
Matrex™	Amicon	No	Yes	45	10
MonoBeads®	Pharmacia/LKB	Yes	No	∼80	10
Nugel™	Separations Ind.	Yes	Yes	30	5, 10
Polywax LP	Poly LC	Yes	Yes	30	5, 7
PL-SAX	Polymer Labs.	Yes	No	100	8, 10
Shim-pack	Shimadzu	No	Yes	30	5
SUPELCOSIL®	Supelco				
SynChropak®	SynChrom	Yes	Yes	30, 100	6.5, 10
TSK® IEX	Toyo Soda	No	Yes	100	5, 10
ZORBAX®	Dupont	Yes	Yes	30	7

a range of technological difficulties. Timely delivery is important and assurances on this matter should be obtained from customer service and other independent users of a product.

2.11 Anion-exchangers

The properties of the ion-exchangers described in Section 2.8 are not all easily obtained or measured. The manufacturer should be asked to supply as much of this information as possible. Once this information has been obtained it will be simpler to predict the type of column needed and in general, some of the chromatographic conditions to be used.

Some of this information is given in *Table 16*, which lists commercially-available anion-exchangers for HPIEC of proteins and polypeptides. Use *Tables 2* and *16* and follow the guidelines in Sections 2.5−2.14 to select the most appropriate ion-exchanger.

2.12 Cation-exchangers

These are employed less commonly than anion-exchangers, since most proteins and polypeptides are acidic and therefore negatively charged (anionic) in the useful pH range 5.5 and above (*Table 5*).

Table 17 lists commercially available columns for cation-exchange chromatography. Again, *Tables 2* and *17* and the guidelines in Sections 2.5−2.14 should be followed in order to select the most appropriate column.

2.13 Mixed- and multi-mode exchangers

A mixed-mode ion-exchanger is defined as a bonded phase which separates components by more than one mechanism simultaneously. In this respect many ion-exchangers for HPIEC are the mixed-mode type, since the mechanism of size-exclusion is generally

Table 17. Commercially available wide-pore HPIEC columns for cation-exchange of proteins and peptides.

Packing	Manufacturer	Strong	Weak	Pore size (nm)	Particle diameter (μm)
AQUAPORE®	Brownlee	No	Yes	30	7
BAKERBOND WIDE-PORE*	J.T.Baker	Yes	Yes	30	5, 15
Chemcosorb	Chemco	Yes	No	30	7
Daltosil, Si	Serva	No	Yes	30, 10	3, 5, 10
MonoBeads®	Pharmacia/LKB	Yes	No	~ 80	10
Nugel™	Separations Ind.	No	Yes	30	5, 10
PolyCAT A™	Poly LC	Yes	Yes	30	6.5
Shim-pack	Shimadzu	No	Yes	30	5
SUPELCOSIL®	Supelco	Yes	No	30	5
SynChropak®	Synchrom	Yes	Yes	30, 100	6.5, 10
TSK® IEX	Toyo Soda	Yes	Yes	100	5, 10
ZORBAX®	Dupont	Yes	Yes	30	7

Table 18. Commercially available wide pore mixed and multi-mode HPIEC columns for proteins and peptides.

Manufacturer	Product name	Application area	Mechanisms
Mixed mode			
J.T.Baker Inc.	BAKERBOND ABx	Monoclonal antibodies	Weak anion- and weak cation-exchange
Alltech Assoc. Inc.	Alltech Mixed-Mode	Acidic peptides	Weak anion-exchange and reversed-phase
Multi-mode			
Rainin	Dynamex-300A AX	Acidic proteins (ion-exchange mode) General proteins (hydrophobic interaction mode)	Weak anion-exchange and hydrophobic interaction

present. In general, however, only interactive mechanisms are considered in determining the separation properties of an ion-exchanger.

Multi-mode ion-exchangers on the other hand are defined as those bonded phases which can be induced to separate the components of a mixture by a different mechanism under different mobile phase conditions. For example a bonded phase may have anion-exchange properties at low ionic strength; but may separate proteins by a hydrophobic interaction mechanism at very high ionic strength. The reason for this is that such a bonded phase will contain positively charged groups and a significant density of hydrophobic groups, all capable of interacting with the surface of a protein.

There are few commercially-available mixed- or multi-mode ion-exchangers. *Table 18* lists most of these bonded phases, whose multiple characteristics are generally emphasized by their manufacturers.

The advantages of these columns are that they are more versatile than single mode columns and that they may exhibit a unique selectivity which may solve many purification problems. A limitation of ion-exchange columns that exhibit hydrophobic character is that tightly bound proteins may not elute at all or at reduced recoveries. As ionic strength increases towards a value at which the protein would elute, the hydrophobic mechanism takes over and the biopolymer is bound more tightly.

2.14 Column use and care

Instructions for column use and care are generally supplied with each HPIEC column. It is worthwhile requesting a copy of these from the manufacturer before purchasing since the instructions usually contain useful information which will help in choosing a column.

The following guideline is necessarily brief and describes the most important aspects of column care. The subject has been reviewed by Wehr (13). Other aspects of this topic are described in Section 2.8.3.

(i)   Examine the column and its container for any signs of damage. The column end caps should be screwed in tight. Do not subject the column to any shock.

(ii)   Check the contents of the column box, which normally contains the column, extra nuts, ferrules and frits, instructions and a QC certificate. Keep the contents inside the box for safekeeping.

(iii)   Read Column Use and Care Instructions.

(iv)   Read the QC certificate carefully and note the column shipping solvent.

(v)   Ensure that the mobile phase in your HPLC is compatible with this solvent.

(vi)   Follow the manufacturers instructions for equilibrating the column. This normally includes a fairly prolonged wash with a low ionic strength buffer at a slow flow-rate. This is followed by a gradient-elution from low to high ionic strength and back.

(vii)   Monitor eluant spectroscopic properties and column back-pressure during the above column conditioning. *Table 2* gives a guide to the typical ranges of pressures you should observe.

(viii) Separate the components of the prepared sample according to the procedure described in Section 2.15.

(ix)   At the conclusion of the work, store the column according to the manufacturers instructions.

(x)   Further extended equilibration is normally unnecessary. It is usually sufficient simply to follow the procedure in Section 2.16, step (xi).

2.15 Starting up: a step-by-step approach to HPIEC

2.15.1 *Separation of three proteins by weak anion-exchange chromatography*

Refer to Appendix II at the end of this chapter for all the items you will need for the initial chromatography. The following procedure assumes that the appropriate buffers and standard protein solutions have been prepared and/or stored according to the guides in Sections 2.2 and 2.3.

(i)   Check the operating condition of your liquid chromatograph. Attach an empty piece of Teflon tubing (0.5 mm i.d., 1.5 mm o.d.) in place of the HPLC column. Remove water-immiscible solvents with isopropanol. Ensure that all wetted surfaces are washed with distilled water (from the mobile phase inlet filters to the outlet tubing of the detector flow cell).

(ii) Replace the distilled water with buffer A using pump designated A. Use a flow-rate of 2 ml per min. Repeat for buffer B using pump B. Buffer A should be the lower ionic strength solution. Run both buffers separately through all wetted surfaces. (Buffers A and B remain separate until the mixer, of course.) Turn the loop valve injector from load to inject and back several times during this process. Then turn it to the inject position.

(iii) Pump buffer A only through the system at 1 ml/min, noting pressure and temperature. Stop the pump. Replace the length of Teflon tubing with a BAKERBOND WIDE-PORE PEI column (4.6 mm × 50 mm) or equivalent column attached at the inlet only.

(iv) Pump buffer A only through the column for 5 min, and collect the effluent (initially the shipping solvent, methanol, followed by buffer A). Examine the effluent for any cloudiness. None should be observed. Stop the pump.

(v) Attach the column outlet to the detector flow-cell and start pump A at 1 ml/min. Observe the pressure which should be between 200 and 300 p.s.i. Monitor the absorbance of the effluent at 254 nm. Measure the pH and the conductivity of the effluent until they match those of buffer A itself. A level baseline should be observed at 254 nm. Continue monitoring pressure of the system whenever the pump or pumps are operating.

(vi) When the column is equilibrated, change the detector wavelength to 280 nm, and the absorbance full scale to 0.1 or thereabouts. Re-zero the baseline.

(vii) Thaw out a working strength solution of cytochrome *c*, and place the sample in a 250 ml beaker containing 50 ml of ice water. Maintain the temperature at or near zero by placing the beaker in a larger bath of wet ice.

(viii) Enter a simple program to carry out a linear gradient from 100% A to 100% B in 2 min at 1 ml/min constant flow-rate.

(ix) Inject 20 μl of the standard protein solution, simultaneously starting the gradient.

(x) Observe the chromatogram as it is formed, for 10 min, noting the maximum absorbance (if possible) of the cytochrome *c* peak if it goes off-scale. If the baseline rises during chromatography more than 20% of the full scale, change the absorbance to 0.2, and double the volume of solution to be injected.

(xi) Reset the program to 100% A and re-equilibrate the column at 2−5 ml/min, depending upon the pressure limits of the HPLC pumps. Equilibrium is reached when the effluent pH matches that of buffer A (about 40−50 ml).

(xii) Repeat steps (ix) and (xi), until chromatography is reproducible. Retention time of cytochrome *c* should be about 0.7 min. Impurity peaks may be observed which should also be reproducible.

(xiii) Thaw out the working strength standards of ovalbumin and β-lactoglobulin A. Chromatograph 20 μl of each protein separately, noting retention times and peak heights in absorbance units.

(xiv) Calculate and note carefully the volume of each standard that will be required to give peaks of equal height in a mixture.

(xv) Prepare and chromatograph the mixture, increasing the volume injected to take into account the dilutions in the standard solutions. Be prepared to use a 500-μl loop.

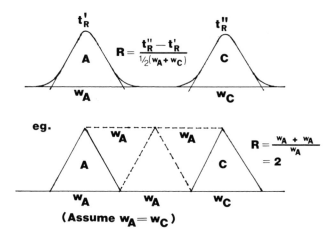

Figure 10. Calculation and meaning of resolution (R) between two peaks.

(xvi) Adjust relative peak heights so that they lie between 30 and 80% of full scale. This may involve increasing the concentration of one or more components in the mixture. Note carefully additions of the extra volumes of protein standards added.

(xvii) The final chromatogram should be close to that shown in *Figure 9* (p. 112), chromatogram A.

2.16 Optimizing resolution

In quantitative terms, the resolution (R) between two peaks in a chromatogram is approximately the number of times either peak can be fitted (at its base) between the peak maxima (*Figure 10*).

The separation of the three proteins described in Section 2.15 can be conveniently carried out under differing eluting conditions to observe the effects of the major parameters upon resolution. These parameters are as follows.

(i)   Buffer types.
(ii)   Ionic strength and pH of buffers A and B.
(iii)   Gradient steepness.
(iv)   Flow-rate.

It is worthwhile to carry out a limited set of experiments to observe the effects of the above parameters on resolution. In real-life samples, similar optimization processes will be important in separating the desired protein or peptide. The following lists the major condition that can be used under the four categories (i) to (iv).

(i) *Buffer types.* Keep buffer A constant (10 mM KH_2PO_4, pH 6.8), and vary buffer B: ammonium acetate (2 M, pH 6.8) + buffer A; ammonium sulphate (2 M, pH 6.8) + buffer A.

(ii) *Ionic strength and buffer pH.* Keep buffer A constant and vary the pH of buffer B. Use pH values of 8, 6 and 5. Change initial ionic strength of elution by mixing

Table 19. Protein mass recovery from BAKERBOND WIDE-PORE ion-exchangers.

Protein	Mass recovery (%)			
	PEI	QUAT*a	CBX*	CARBOXY-SULFON*a
Bovine serum albumin	97	95	96	95
Ovalbumin	99	97	98	97
Transferrin	97	99	100	99
Haemoglobin	97	95	95	96
Myoglobin	97	98	99	98
Cytochrome c	99	95	98	97
Lysozyme	96	98	97	95
Carbonic anhydrase	96	92	98	97
β-Lactoglobulin	98	100	100	98
Human serum IgG	97	95	96	96

aCurrently in development.

buffer B (pH 6.8) at percentages of 10, 20 and 30. Run gradients to 100% B, keeping gradient slope constant.

(iii) *Gradient steepness.* Run linear gradients of 5, 20 and 60 min.

(iv) *Flow-rate.* Run the standard 2 min gradient using flow-rates of 0.3, 2 and 5 ml per min.

Calculate and tabulate values of R for the ovalbumin and β-lactoglobulin peaks, separated under the various conditions described above. These will provide a first-hand idea of the various controls that can be exerted over resolution in HPIEC.

In general, it should be observed that resolution will increase as the following changes occur:

(i)   as buffer B changes from KH_2PO_4 to $(NH_4)_2SO_4$ to NH_4 acetate;

(ii)   as the pH of buffer B decreases;

(iii)   as the ionic strength of buffer A increases (up to a maximum);

(iv)   as the gradient steepness decreases; and

(v)   as the flow-rate decreases.

2.17 **Measuring recoveries of mass and activity**

Mass recovery is usually measured as the weight percentage of an injected peptide or protein that is eluted from the column. Modern, high-performance columns for HPIEC should give 90−100% recoveries of a wide variety of proteins (*Table 19*). Recovery may depend upon the nature of the column packing, the quantity of protein injected, the history of the column and the general chromatographic conditions employed (Section 2.15).

The simplest procedure used to measure recovery is as follows.

(i)   Inject 1−5 mg of the peptide or protein of interest onto the column and chromatograph according to the method of Section 2.15. Collect the entire peak corresponding to the eluted protein in a volumetric flask (10 ml). Make up to the mark with the buffer as it elutes from the column. Stop the pump.

(ii)   Disconnect the column from the injector, and inject the same quantity of protein through the loop and directly into a volumetric flask (10 ml). Make up to the mark with buffer from the pump.

(iii)   Repeat step (i).

(iv)   Measure the absorbance of the standard and eluted protein solutions using a spec-
         trophotometer set at a wavelength where absorbance can be measured accurately.
         Ensure that neither buffer A nor buffer B absorb strongly at this wavelength.

(v)   Calculate the percent recovery by dividing the measured absorbance of the
        non-chromatographed protein solution into that of the chromatographed protein
        solution, and multiplying by 100. Repeat for the duplicate, which should be within
        10% of the first measurement.

Other methods of measuring mass recovery are in common use, such as radiolabelling,
re-injecting the collected peak and area ratioing using internal standards.

Activity recoveries are expressed as the percentage of activity of the peptide or protein
observed before chromatography that is recovered after chromatography. This parameter
must take into account the same recovery of the active polymer. For example if only
50% of the mass of the protein is recovered, this will cause a loss of 50% in activity.

The technique of measuring activity recovery is more complex than that of mass
recovery, and its description will not be attempted in this chapter. The reader is referred
to several detailed methods in the literature (14,15) for measuring activity of
biopolymers. Many kits for measuring activity of certain enzymes are commercially
available (16), which contain detailed practical instructions.

2.18 Collecting and storing fractions

Components of the separated mixture may be collected for further analysis or processing.
This may be done automatically or manually.

The details of operation of a fraction collector will not be described in this chapter,
and should be obtained from the appropriate manual. Manual collection is simpler,
especially for non-routine separations, but demands close attention to the recorder which
monitors the protein bands as they pass through the flow-cell.

The manual collection of protein and/or polypeptide fractions can be carried out
according to the following steps.

(i)   Ensure that the liquid chromatograph is in full working order. Buffers and samples
        should be ready for use.

(ii)   Attach a fine-bore (0.010 inch) piece of Teflon tubing (1/16 inch o.d.), about
        0.5 m long to the flow-cell outlet. Calculate its internal volume in millilitres.
        Place its outlet in a conical flask. This tubing will serve as both a back-pressure
        device and a convenient directional tube, which is placed in separate containers
        as each fraction is collected.

(iii)   Place a labelled test tube (13 mm × 100 mm, glass) rack into ice water in a
        polystyrene container.

(iv)   Mark 20 or more test tubes in water-insoluble ink, using a code to indicate the
        following pieces of information: fraction number, sample identification, run
        number.

(v)   Place the test tubes in the rack in a sequence that will make it convenient and
        accurate to collect fractions. Allow them to cool to ice water temperature.

(vi)   Set the recorder chart speed to run 4−5 times as fast as normal. This will make

Table 20. Applications of HPIEC to the separation of mixtures of macromolecules.

Analyte	*Reference*
Alkaline phosphatase	38
Apolipoproteins	18
Basic peptides	19
Calmodulin	20
Creatine kinase	22
General strategies	35, 44
Growth factors	25
Haemoglobins	24, 33, 34, 40, 43
Hormones	30
Insect venoms	36
Monoclonal antibodies	26, 27, 41
Plant globulins	29
Proteins - (i) General	17, 32, 42, 44
(ii) Specific	
Cereal	21
Complex	32
Membrane	31
Red cell membrane	39
Ribosomal	28
Serum	23
Viral	37
Sample preparation	6

        it easier to mark the fraction position on the chromatogram. Place a watch in view.

(vii)   Begin the chromatography. If the separation has not been carried out before, ensure that all peaks are collected. Since each part of all the fractions will elute some seconds after it appears in the flow-cell, this must be taken into account. As soon as a peak forms in the flow-cell, place a mark on the chromatogram, and wait for the calculated time to elapse before collecting the peak. The calculated time is obtained by dividing the volume of tubing attached to a flow-cell by the flow-rate.

(viii) When no more components elute from the column, seal each test tube with Parafilm, and place the entire rack into a refrigerator. Further tests on each fraction (such as electrophoresis or activity assay) should be performed as soon as possible.

(ix)   It is most important to keep together all data, chromatographic conditions and noted locations of labelled fractions.

3. APPLICATIONS

The technique of HPIEC of proteins and polypeptides has developed at an accelerating pace since the first applications in 1976. The number of applications is extraordinarily large, and the list given in *Table 20* is consequently very selective. The references given in this table describe more or less detailed procedures and results in HPIEC of proteins and polypeptides. They cover a range of applications of varying complexity, several classes of proteins and polypeptides and illustrate the use of the major commercially-available columns. *Figures 1, 2, 3* and *4* illustrate the HPIEC of a number of naturally occurring sources of proteins. The set of chromatographic conditions given with each figure is an accurate guide to obtaining reproducible chromatography with the respective samples. Sample preparation is important for protein-containing samples, and guidelines for this technique are given in Section 2.4.

4. ACKNOWLEDGEMENTS

I would like to thank Drs Laura J.Crane and Harold A.Kaufman for their practical comments on this manuscript; Drs Steven A.Berkowitz and David R.Nau for the chromatography; Ms Joanne Volkert and Susan Kurasz for the excellent production of this chapter.

5. REFERENCES

1. Rounds,M.A. and Regnier,F.E. (1984) *J. Chromatogr.*, **283**, 37.
2. Drager,R.R. and Regnier,F.E. (1986) *J. Chromatogr.*, **359**, 147.
3. *LC.GC Magazine* (1987) Buyers' Guide.
4. American Laboratory (1988) Buyers' Guide Edition.
5. McNair,H.M. (1987) *J. Chrom. Sci.*, **25**, 564.
6. Wehr,C.T (1987) *J. Chromatogr.*, **418**, 27.
7. Wehr,C.T. and Majors,R.E. (1987) *LC-GC Magazine*, **5**, 548.
8. Suelter,C.H. (1985) *A Practical Guide to Enzymology*. John Wiley and Sons, Inc., New York.
9. Brewer,J.M., Pesce,A.J. and Ashworth,R.B. (1974) *Experimental Techniques in Biochemistry*. Prentice-Hall, New Jersey.
10. *Methods in Enzymology* (1986) Cumulative Subject Index, Academic Press, Inc., Orlando, Florida, Vols. 81–94, 96–101.
11. LIPOCLEAN^{TM} is a product of Hoechst Behringwerke AG.
12. Chicz,R.M., Shi,Z. and Regnier,F.E. (1986) *J. Chromatogr.*, **359**, 121.
13. Wehr,C.T. (1984) In *Methods in Enzymology*, Jakoby,W.B. (ed.), Academic Press, New York, Vol. 104, Part C, 133–169.
14. Bernhard,S.A. (1968) *The Structure and Function of Enzymes*. W.A.Benjamin, Inc., CA, Chapters 4 and 8.
15. Barman,T.E. (1969) *Enzyme Handbook*. Springer-Verlag, New York, NY, Volumes 1 and 2.
16. See Appendix III.
17. Chang,S.H., Gooding,K.M. and Regnier,F.E. (1976) *J. Chromatogr.*, **125**, 103.
18. Ott,G.S. and Shore,V.G. (1982) *J. Chromatogr.*, **231**, 1.
19. Cachia,P.J., VanEyk,J., Chang,P.C.S., Taneja,A. and Hodges,R.S. (1983) *J. Chromatogr.*, **266**, 651.
20. Berkowitz,S.A. (1987) *Anal. Biochem.*, **164**, 254.
21. Bietz,J.A. (1985) *Cereal Chem.*, **62**, 201.
22. Wux,A.H.B. and Gornet,T.G. (1985) *Clin. Chem.*, **31**, 25.
23. Schlabach,T.D. and Abbott,S.R. (1980) *Clin. Chem.*, **26**, 1504.
24. Hanash,S.M. and Shapiro,D.N. (1981) *Hemoglobin*, **5**, 165.
25. Sullivan,R.C., Shing,Y.W., D'Amore,P.A. and Klagsbrun,M. (1983) *J. Chromatogr.*, **266**, 301.
26. Nau,D. (1986) *Biochromatography*, **1**, 82.
27. Crane,L. (1987) In *Monoclonal Antibody Production Techniques and Applications*. Schook,L.B. (ed.), Marcel Dekker, Inc., New York, Chapter 9.
28. Capel,M., Datta,D., Nierras,D.R. and Craven,G.R. (1986) *Anal. Biochem.*, **158**, 179.
29. Lambert,N., Plumb,G.W. and Wright,D.J. (1987) *J. Chromatogr.*, **402**, 159.
30. Pearson,J.D., McCroskey,M.C. and DeWald,D.B. (1987) *J. Chromatogr.*, **418**, 245.
31. Welling,G.W., VanDerZee,R. and Welling-Wester,S. (1987) *J. Chromatogr.*, **418**, 223.
32. Regnier,F.E. (1987) *J. Chromatogr.*, **418**, 115.
33. Huisman,T.H.J. (1987) *J. Chromatogr.*, **418**, 277.
34. Wilson,J.B. and Husiman,T.H.J. (1986) In *The Hemoglobinopathies*. Huisman,H.T.J. (ed.), Churchill Livingstone, Edinburgh.
35. Hearn,M.T.W. (1987) *J. Chromatogr.*, **418**, 3.
36. Einarsson,R. and Renck,B. (1984) *Toxicon*, **22**, 154.
37. Welling,G.W. (1984) *J. Chromatogr.*, **297**, 101.

38. Britton,V.J. (1983) *Liq. Chromatogr.*, **1**, 176.
39. Lundahl,P. (1984) *J. Chromatogr.*, **297**, 129.
40. Ip,C.Y. and Asakura,T. (1986) *Anal. Biochem.*, **156**, 348.
41. Dorfman,P. (1984) *Gen. Eng. News*, **May/June**, 15.
42. Alpert,A.J. (1983) *J. Chromatogr.*, **266**, 23.
43. Ou,C.N., Buffone,G.J. and Alpert,A.J. (1983) *J. Chromatogr.*, **266**, 197.
44. Berkowitz,S.A., Henry,M.P., Nau,D.R. and Crane,L.J. (1987) *Amer. Lab.*, **May**, 33.

6. APPENDIX I

HPLC bibliography

1. *Applications of High Speed Liquid Chromatography*, by Done,J.N., Knox,J.H. and Loheac,J., published by Wiley, 1974.
2. *High-Performance Liquid Chromatography of Proteins and Peptides: Proceedings of the First International Symposium*, edited by Hearn,M.T.W., Regnier,F. and Wehr,C.T., published by Academic Press, 1983.
3. *High-Performance Liquid Chromatography: Advances and Perspectives*, Volumes I to IV, edited by Horvath,C., published by Academic Press, 1980.
4. *Instrumentation for High Performance Liquid Chromatography*, edited by Huber,J.F.K., published by Elsevier, 1978.
5. *Instrumental Liquid Chromatography: A Practical Manual on HPLC Methods*, by Parris,N.A., published by Elsevier, 1976.
6. *Practical High Performance Liquid Chromatography*, edited by Simpson,C.F., published by Heyden, 1976.
7. *Introduction to Modern Liquid Chromatography*, 2nd edition, by Snyder,L.R. and Kirkland,J.J., published by Wiley, 1979.

7. APPENDIX II

High-performance ion-exchange chromatography of cytochrome *c*, ovalbumin and β-lactoglobulin A: materials and equipment

Equipment

1. A high-performance liquid chromatograph from any suppliers. Must be capable of gradient-elution, flow-rates from 0.1 to 10 ml per min, at pressures up to 3000 p.s.i. Requires detection at 254 and 280 nm. Dual pen recorder. Injector loops of 50, 100 and 500 μl vol.
2. Balance for weighing in milligrams.
3. Culture tubes, 13 × 100 mm, borosilicate glass.
4. Test-tube rack for culture tubes.
5. Insulated container to hold test-tube rack.
6. Refrigerator with freezer.
7. pH meter and standard solutions for calibration.
8. Flow-through pH meter, attached to outlet of detector's flow-cell (optional, but very useful).
9. General glassware and plastic ware [beakers, pipettes, squeeze bottles, Pasteur pipettes, Parafilm, side-arm filter flask (2−4 litre)].
10. Ultrasonic bath.
11. Magnetic stirrer/hot plate.
12. Source of vacuum.
13. Eppendorf or Oxford pipettes.
14. Eppendorf or similar microcentrifuge.
15. Filter holder and membranes (0.45 micron).
16. Narrow necked, polyethylene, 2 or 4 litre containers for buffers.
17. Vortex mixer.

Chemical and protein standards

1. Potassium phosphate (monobasic, HPLC grade.
2. HPLC grade water or deionized water (4 litres).

124

3. Cytochrome *c*.
4. Ovalbumin (hen egg white).
5. β-Lactoglobulin A.
6. Potassium hydroxide solution, 1 M.

HPLC column

1. BAKERBOND WIDE-PORE PEI, 5 micron, 4.6 mm × 5 cm, ion-exchange HPLC column (PEI stands for polyethyleneimine covalently bound to porous silica) or a similar, short anion-exchange column specifically designed for protein chromatography.
2. FPLC adaptor: for users of the Pharmacia FPLC system. allows the above HPLC column to be fitted directly to the FPLC injector and detector.

8. APPENDIX III

Commercial sources of purified proteins and peptides

1. Sigma Chemical Company, St. Louis, MI, USA.
2. United States Biochemical Corporation, Cleveland, OH, USA.
3. Serva Fine Biochemicals, Heidelburg, West Germany.
4. ICN Biochemicals, Cleveland, OH, USA.

CHAPTER 5

Reversed-phase chromatography of proteins

P.H.CORRAN

1. INTRODUCTION

There are signs that the reversed-phase (RP) chromatography of proteins may have passed the zenith of its popularity, overtaken by the more recent development of high efficiency ion-exchange and hydrophobic interaction supports which can deliver separations comparable to RP-HPLC in resolution, but carrying greatly reduced risks of denaturation and loss of biological activity. Nevertheless there are many circumstances in which denaturation may be unimportant, such as in structural studies, or where structure and activity are retained despite the high concentrations of organic modifier inevitably involved and where a rapid and powerful separation technique of unusual specificity like RP-HPLC is of great service.

1.1 Status of technique

The discovery that RP chromatography could be applied to proteins as well as peptides was made nearly 10 years ago. For some time attempts to develop useful separations employed the same packings as were used for simple molecules, with variable success. As the requirements for the successful chromatography of proteins have been formulated, manufacturers have responded by producing ranges of RP packings designed for use with macromolecules and a number of fairly standardized sets of elution conditions have come to dominate published separations. The very simplicity of RP chromatography has made it easy for workers to tailor conditions to suit their own particular problems. This facility has been important since the characteristics of a protein may play a large part in how easy it is to recover: it is not possible to give a set of separation conditions to suit all or even most situations. It is possible, however, to suggest a starting point for a separation and to place the process of optimizing this separation on a basis which is, at least, only semi-empirical.

1.2 Uses

RP chromatography has been used to separate a wide range of proteins of many types, but much of the published work has been concerned with a few restricted classes: growth factors, peptide hormones and immunomodulators, virus proteins, haemoglobins and ribosomal proteins account for half of published separations. It is no coincidence that, in most of these cases, the proteins were either structurally robust or the retention of biological activity was of secondary importance.

1.3 **Hydrophobic interaction chromatography (HIC) and reversed-phase chromatography (RPC)**

Chromatographic supports for both RPC and HIC usually consist of hydrophobic groups covalently attached to a hydrophilic matrix. The difference lies in the density of these groups, which, in HIC, are relatively widely spaced and designed to interact with hydrophobic patches or pockets on the surface of an intact folded protein, but in RPC are densely attached to cover as much of the surface as possible so as to provide a hydrophobic surface, perhaps with a layer of the organic component of the mobile phase, to which the protein adsorbs. These conditions of RPC tend to trap any unfolded protein present in the unfolded, more strongly-bound state. Retention in HIC thus depends on distinct structural features formed by the way in which the native protein folds, while RPC depends much more on the overall amino acid composition, though structural features also play an important though less well defined part in the retention mechanism.

1.4 **Format and scope of chapter**

This chapter deals with the application of RPC to proteins in six main sections. For the purpose of convenience a protein is assumed to be insulin and anything larger than insulin (mol. wt 5600), but excluding fragments of larger proteins. The first (Section 2) sketches out some of the theoretical background to the HPLC of macromolecules: this is of importance because of its implications for the choice of chromatographic supports and in understanding at least some of the reasons why proteins do not always behave in a manner predictable from experience with smaller molecules. This in turn leads on to a brief consideration of how proteins interact with reversed-phase packings, and how this may affect chromatographic behaviour. The next three sections deal with the nuts and bolts of a chromatographic separation: apparatus, reversed-phase packings, suitable column sizes, mobile phases and detection systems. Sections 6 and 7 cover the application of this to a chromatographic separation: how to modify a separation, possible ways of improving resolution, maintaining activity or dealing with awkward molecules. The final section describes some examples of protein separations, chosen to illustrate some of the points made earlier.

2. THEORETICAL CONSIDERATIONS

2.1 **Chromatographic behaviour of large molecules**

The behaviour of a macromolecule during chromatography is not always obvious when approached from an experience with small molecules. In the case of RPC two particularly striking observations are that the elution of a protein from a reversed-phase column is extremely sensitive to the smallest variations in the concentration of the organic modifier—a 1% change in the proportion of acetonitrile or propanol may be the difference between complete retention and almost immediate elution—and that extremely poor protein separations may occur on columns which exhibit high efficiencies when tested with small solutes, while good separations may be obtained on columns which appear very inefficient under test conditions. The first is a manifestation of a general property of a large molecule and the second is not necessarily inconsistent with the general theoretical principles governing chromatographic behaviour. The discussion that follows is an outline; further details can be obtained through refs 1 and 2.

2.1.1 *Dependence of retention on modifier concentration*

It is a general and obvious feature of the adsorption of a large molecule, by whatever mechanism (ion-exchange or hydrophobic interaction), that the extent of interaction, measured by the number of charged or hydrophobic groups involved, is greater than for a smaller molecule. The effect of this in RPC can perhaps be most simply explained by considering the effect of a change of modifier concentration on the retention of a simple molecule—for example, an amino acid—and comparing it with that of a protein. When the concentration of the organic solvent component of the mobile phase is changed, the free energy of transfer of a single hydrophobic amino acid side chain from the bulk mobile phase to the surface of the adsorbent will change by an increment δG. In the case of a peptide, a similar adsorption step might involve the transfer of several such hydrophobic side chains, say 10, with a change in free energy of transfer of $10 \times \delta G$. When a protein is involved the number of hydrophobic residues involved can be much greater, say 100, and the change is $100 \times \delta G$. Strictly the term 'hydrophobic' is something of a misnomer, since the free energy gain on transfer from solution to the non-polar surface is provided by the interactions the protein was preventing the solvent molecules making with each other—as Tanford notes (3) it is the water which dislikes the solute, not vice versa. A measure of the strength of solvent – solvent interations is provided by the surface tension and a measure of the number of such interactions prevented is given by the surface area of the hydrophobic region, which accounts for the importance of these two figures in theoretical treatments. Since the relationship between the equilibrium constant for the distribution between free solution and the adsorbent surface, k, and the free energy of transfer, $\triangle G$, is governed by the familiar equation:

$$\triangle G = RT \ln k$$

it is clear that the longer the peptide chain, the more hydrophobic side chains will be involved in transfer and the more steeply peptide retention is likely to depend on modifier concentration. The actual value of the equilibrium constant depends not only on the free energy contributions from the transfer of hydrophobic residues to the environment of the adsorbent surface, but also on the opposite contributions from the transfer of hydrophilic groups. The steepness of its dependence on mobile phase composition depends on how many solvent – solvent interactions are prevented by each molecule of solute in solution; a small change for a few solvent molecules rapidly adds up to a considerable difference when summed over many.

The direct consequence is that, even in the absence of other constraints (see Section 2.2), it is much more difficult to carry out reproducible isocratic separations of proteins and it is usual practice to use gradient-elution except in certain specialized and favourable cases: other possible complicating factors are discussed below.

2.1.2 *Chromatography efficiency*

Column efficiency is best given in terms of the reduced plate height, h, and the reduced velocity, v, at which the column is operated. The simplest form of the equation governing the efficiency of a chromatographic column (as measured in these terms) is the dimensionless version proposed by Knox (4) summarized below.

The Knox equation. This equation was derived following detailed consideration of the various band dispersion mechanisms which operate in chromatography under isocratic conditions. Knox decided that there were three dispersion processes operating simultaneously and that the contribution of each to the reduced plate height was additive:

$$h = h_{(\text{axial diffusion})} + h_{(\text{flow})} + h_{(\text{mass transfer})}$$

By calculating each of the individual contributions Knox obtained his equation which relates the measured reduced plate height and the reduced velocity as follows.

$$h = Av\ 1/3 + B/v + Cv$$

Where H = plate height, d_p = particle diameter of the packing, u = linear velocity of the mobile phase (column length/to), D_m = diffusion constant, then $h = H/d_p$ = reduced plate height and $v = ud_p/D_m$ = reduced velocity.

This equation passes through a minimum which, for a small molecule and a well packed column, is found to occur at a value for h of about 2. For a conventional 4.6 mm i.d. column filled with a 5 micron reversed-phase packing in methanol/water mixtures the flow-rate of this minimum is about 0.5 ml/min, and v of the order $3-5$. The A term describes the contribution to dispersion due to tortuous flow through a packed bed, and is a measure of how well the column is packed. The B term describes the effects of axial diffusion (in the direction of flow) on dispersion: at normal flow-rates the influence of this term is relatively unimportant. The C term is concerned with the effects of slow equilibration between the mobile and stationary regions due to, for instance, the diffusion of solute into and out of the particle and the kinetics of adsorption and desorption. For a well packed column and a small solute $A \leq 1$, $B \simeq 2$ and $C \leq 0.1$.

Under normal conditions of use with normal small particle support (5 μm) packings and with small molecules being separated v is quite close to the optimum—in the region of 10, and the value of h is determined largely by the value of the A term of the equation. For a protein D_m may be 10 times larger than for a small molecule, v more like 100, and h is determined more by the C term. In other words what matters most for a small molecule is that the column should be well packed, but for a protein it is far more important that diffusion within the pores of the packing and interaction with the packing should be as rapid as possible. Typical traditional RP packings are based on silicas with mean pore diameters of $6-12$ nm and pore size distributions wide compared to the narrow pore size distributions encountered in the traditional gel filtration media used in biochemistry (see Chapters 1 and 3). Typical maximum dimensions of globular proteins range from about 3 nm for cytochrome c (mol. wt 12 600) to 10 nm for alcohol liver dehydrogenase (mol. wt 80 000) or catalase (mol. wt 250 000) and very much greater for fibrous proteins such as collagen. The restricted freedom of diffusion within the pores of the packing results in a higher value of C than for a small molecule and exclusion from the pores to a reduction in the surface area available for adsorption. It is thus advisable—if not essential with large proteins—to use packings with larger mean pore diameters; 30 nm is a popular size.

2.1.3 *Gradient-elution*

When these relations are considered in the context of gradient- rather than isocratic-elution there are some important additional consequences (1): the resolution of a gradient separation of a mixture of macromolecules depends directly on the particle diameter of the packing, but only weakly on the column length and flow-rate for gradients of equal time. However, the peak height, which determines the detection limit, depends inversely on the particle diameter, flow-rate and, less strongly, on the gradient time. An increase in the length of the column has little effect on resolution since by the time the protein band reaches the extra column length it has been overtaken by the gradient and the modifier concentration is sufficient to ensure that the protein spends very little time in the adsorbed state; the effect of the additional length is thus largely passive. There is thus no point in using long columns for protein separations except for isocratic work. If extra capacity is needed one might just as well use a short fat column as a long thin one. An increase in the flow-rate reduces the effective plate height but is compensated for by the improved resolution resulting from a gradient which is shallower in volume terms. Theoretically, therefore, to reduce the peak volume and improve resolution one should use a shallower gradient and a lower flow-rate; this is indeed often observed (5) and would certainly be the case for a 'well-behaved' macromolecule. The next section summarizes some of the reasons why proteins are not always ideal solutes in this sense.

2.2 The interaction of proteins with RP packings

A second group of observations of protein behaviour during RP-HPLC concerns peak shapes, multiple peaks and variable recoveries of material which may depend on column loading or the steepness of the gradient used. A protein which appears to behave satisfactorily in gradient-elution may be impossible to chromatograph isocratically at modifier concentrations close to the concentration at which it was eluted in the gradient. Protein may not be completely eluted and may give rise to peaks on subsequent blank gradients. On the other hand a protein may not be retained at all at neutral pH, only under acidic conditions. Some of these phenomena—poor recoveries of very small amounts of protein and some kinds of peak tailing—can be laid at the door of the packing in use. Others however, like fronting, elution of the same protein in multiple peaks and poor recoveries of moderate or high protein loads are most plausibly explained in terms of protein structure.

2.2.1 *Structure of proteins in RPC mobile phases*

A globular protein normally has a stable 'native' conformation in neutral aqueous solution. Usually this structure is an equilibrium one and folding and unfolding behave as simple two state processes without intermediate stages, at least on a chromatographic time scale. For ribonuclease A, for instance, the first order unfolding rate constant

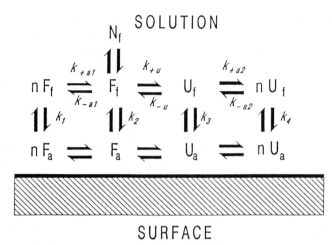

Figure 1. A simple model of protein/support interactions. The protein is assumed to consist of a partly folded state, F, in equilibrium with the native conformer, N_f, an aggregated form, nF, and an unfolded state, U, which in turn is in equilibrium with a second aggregated form, nU. It is assumed that the protein must at least partially unfold to interact with the packing. The interconversions between these different states are determined by the rate constants shown, and the relative amounts at equilibrium by the ratios of the forward and reverse rate constants (the equilibrium constants). Each form is in turn distributed between mobile phase and sorbent with equilibrium constant $K_1 \ldots K_4$. The time constant, τ is the reciprocal of the rate constant, so: $\tau_{+u} = 1/K_{+u}$ and $\tau_{-u} = 1/K_{-u}$.

is about 5×10^{-4}/sec at 40°C and pH 8 (time constant, $\tau = 2000$ sec). Under these conditions there must be, perhaps, 0.1% of unfolded molecules present at any time (6). The status of a protein which normally operates in an environment such as a membrane is less well defined and occasionally the native state is kinetically determined in that the native structure is thermodynamically unstable in 'native' environments, but the rate constant of denaturation is very small: something of this kind occurs with the glycoprotein hormones (7). Agents such as urea, guanidine, propanol, acetonitrile or trifluoroacetic acid may accelerate denaturation, as can a change of pH or temperature, and they may also act to keep the denatured protein in solution, for instance by reducing the tendency to aggregate. There is evidence that proteins may interact with the organic solvents used in RPC in a much more specific way (8). The typical conditions used in RPC of peptides and proteins—low pH and high concentrations of organic solvents thus favour denaturation, so that a protein dissolved in mobile phase will consist of a mixture of native, partly folded, unfolded and aggregated structures. The proportions of the different components of this mixture will be determined by the kinetic and equilibrium characteristics of the transitions between the different forms, and will be influenced by the composition of the solvent. Any account of the chromatographic behaviour of a protein thus has to take account of the different folded structures it may assume and the rapidity with which they are in equilibrium.

2.2.2 *Interaction with reversed-phase packings*

There are several possible models of the retention and desorption of a solute by a reversed-phase surface (2,8). For practical purposes it seems to make little difference at present which one considers. Whatever the mechanism, the interaction between the surface of the packing and the protein depends on the equilibria and constituent rate constants of the different protein structures present. A simplified scheme for the equilibria involved is set out in *Figure 1* (9,10), though in fact there is evidence for the involvement of more than one denatured form in model proteins (8,11) resulting from a rapid change associated with initial adsorption or from a slower change which may occur once on the surface. The evidence available (9,11) supports the assumption that denatured structures are more retained than native ones ($k_3 > k_2$) and it seems a natural assumption from the fact that many proteins are not retained at neutral pH that some form of unfolding must occur often for the protein to bind to the surface of the support. For instance if RNase A was injected onto a RP column with $t_0 = 4$ min with a first-order rate constant for unfolding of 5×10^{-4} per sec (see previous section) and if the native form ('N' in *Figure 1*) is not adsorbed but the partly unfolded form ('F_f') is, then 11% of the RNase would unfold during this time and be available for adsorption while 89% would remain folded and pass straight through the column. Aggregated forms may range from a 'precipitate' on the packing surface (nU_a) to a more defined association (nF) such as is perhaps found with the glycoprotein hormones. If the equilibria are rapid, then the predominant form retained on the surface will be the one with the tightest binding constant. This appears to be the case with ribonuclease A, which exists on the surface entirely as an unfolded form (U_a in *Figure 1*), but which is able to re-fold at a significant rate once its concentration in the bulk liquid phase (U_f) reaches significant levels (9). In the case of the glycoprotein hormone thyrotropin, however, it is possible to choose chromatographic conditions such that the native dimer (nF_f) does not significantly dissociate during chromatography (k_{+al} small). It must therefore be bound to the support as the dimer also. For this case an extra species, nN_f, is necessary in *Figure 1*.

2.2.3 *Effect of chromatography*

For a system such as that illustrated in *Figure 1* in rapid equilibrium, no effect on peak shape would be observed and the effective capacity constant would be determined by the largest of $k_1 \ldots k_4$, since all forms would rapidly finish up bound as the structure with greatest affinity. If any of the time constants involved is of comparable magnitude to the peak width then the peak shape may be affected. This effect is greatest when the time constant is of the order of the peak width, though the exact effect depends on the relative retentions of the species involved. For instance, if $k_3 > k_2$, the predominant form bound is unfolded, U_a and k_{-u} corresponds to a time constant less than but of the same general magnitude as the peak width then the result will be fronting, the severity of which depends on the size of k_{-u}, the average concentration of U_f during elution and the steepness of the gradient, which determines how long significant concentrations of U_f exist during elution. This appears to be the case for RNase in isopropanol gradients at pH 4 (9), where a full explanation of the observed behaviour needs to consider also the effect of isopropanol concentration on k_{-u}. An alternative case, where $k_3 > k_2$ and the predominant form bound is F_a might present as a situation where recoveries of F_f depended on the steepness of the gradient, which determines the length of time significant concentrations of F_f persist. The unfolded form might be eluted as a peak later in the gradient, or might not be eluted at all until the lower organic modifier concentrations in the earlier part of the gradient allowed a proportion to re-fold (perhaps on the surface) and elute as a peak in a later chromatogram. Poorer recoveries at high loadings than at moderate ones might be due to aggregation (10). Some slow steps in protein re-folding are known to be associated with proline isomerization, for instance Pro93 in RNase A (12), and it is actually possible to elute the two isomers of a proline peptide as separate peaks (13); apparent re-folding rates may thus depend on how long a protein has spent in a denatured state.

2.3 Practical conclusions

The simplified schemes outlined in the previous sub-sections suggest that, at least in part, the anomalous behaviour of proteins in RPC can be placed within a rational framework. Unfortunately a major component of this framework is the poorly understood mechanism of protein folding, and the equally poorly understood subject of the interaction of a protein conformation with a reversed-phase surface. For peptides up to a length of 30 residues or so it is possible to predict retention with reasonable accuracy on the basis of the amino acid composition, provided such complicating factors as disulphide bridges are set to one side. These prediction methods do not give accurate results for longer sequences, presumably due to the tendency to form stable folded structures. So although the general strategy for chromatographing a protein remains empirical, it is possible to make certain rational choices based on the desired outcome.

3. APPARATUS

The apparatus needed for the RP-HPLC of proteins need not differ from that used with smaller molecules. Some points affecting choice of apparatus are set out below.

3.1 Pumping system

Most commercial gradient systems are suitable for protein separations. The only points needing consideration are firstly, if it is necessary to carry out a separation at a low flow-rate then it is desirable to minimize the dead volume between the point at which the gradient is mixed and the injection point—5 ml at 0.2 ml/min represents a lag of 25 min. Secondly, under such circumstances the pumping system must be capable of delivering a small proportion of either solvent accurately and reproducibly. If a twin pump system is employed then at a flow-rate of 0.2 ml/min, 5% of one solvent represents a flow-rate for that pump of 10 μl/min. Many commercial machines are now capable of delivering this kind of performance. Occasionally extra active mixing may need to be added when viscous propanol-containing mobile phases are used for gradient generation. Some of the solvent systems described later contain HCl and NaCl which may corrode stainless steel. If prolonged use is to be made of halide containing systems, stainless steel apparatus can be 'passivated' by flushing with 2 M nitric acid overnight (without column or pre-column!), a process which confers partial protection, but it is wise to discuss this with the manufacturer also. For occasional use it is sufficient to flush the system with water each evening. Alternatively there are now several HPLC systems available which contain no exposed stainless steel parts.

3.2 Injection of sample

For most purposes a conventional HPLC injection valve is entirely adequate. If it is necessary to apply a large volume of dilute sample a convenient way is to pump the sample as one of the limiting solvents. Preferably the hold-up volume of the pump should not be too large. Two alternatives are to use a cheap auxiliary pump connected through a low-volume tee or to invest in one of the large-volume all glass injection devices sold by Pharmacia and accept the pressure limitations.

3.3 Column hardware

Conventional columns are constructed from stainless steel components. Although proteins may interact with metal surfaces such as column walls and frits, considerations of surface area and transit time suggest that a protein molecule has little opportunity to do so during passage through a conventional column. Virtually all interaction is with the column packing, which has an area of tens or hundreds of m^2 rather than the tens of cm^2 of the column wall, and with which the protein is in contact for minutes or tens of minutes rather than the fraction of a second taken to traverse a frit. This is to put matters in perspective rather than to suggest that losses due to interaction with stainless steel never occur. If such losses on metal surfaces are found the most likely candidates are the entry and exit frits or meshes or the connecting tubing, and a simple solution may be to replace these with inert polymer or glass fibre equivalents. Corrosion in halide containing mobile phases is much more of a problem in reversed-phase work. Several suppliers now offer glass walled columns with non-metallic or non-stainless fittings. These are often part of a cartridge system, which may offer attractive savings in costs and the loss of robustness is unimportant since short columns operated at modest pressures give excellent results. The drawback at present is that the range of packings available in glass or glass-lined columns is severely limited.

3.4 Detectors

(i) By far the most common detector is the variable wavelength UV monitor operated at $210-220$ or at 280 nm. The old-fashioned fixed wavelength 254 nm monitors are less suitable for protein detection. Since macromolecules are eluted as fairly broad peaks (typically $2-5$ ml for conventional columns) they place modest demands on the dimensions of the flow cell. A conventional 1 mm i.d. 8 μl flow-cell can be replaced by a 50 μl 2.5 mm i.d. cell to improve transmission with opaque solvents without significant loss of performance. However, it is as well to be aware of the possibility of stray light artefacts in these circumstances.

(ii) A useful, but surprisingly little used, alternative to UV absorption is to monitor the intrinsic tryptophan fluorescence. The usual excitation wavelength is 280 nm, but additional sensitivity is gained by exciting at a wavelength in the region $210-225$ nm. Emission is monitored at about 340 nm in both cases. The advantages are a potential several-fold increase in sensitivity and an improvement in the selectivity of detection, particularly in distinguishing protein peaks from 'ghost' peaks due to impurities in solvents. The disadvantage is that proteins which do not contain tryptophan are not detected with comparable sensitivity.

(iii)   The second most popular direct detection system is to divert a portion (5 – 10%) of the effluent into a continuous flow system for reaction with fluorescamine. Amino groups, but not ammonia, give a fluorescent product which is monitored in a suitable fluorimeter. Apart from the added complexity, the only disadvantage is in the unlikely event that a protein possesses no free amino groups. The system is potentially more sensitive than UV absorbance and is essential for the pyridine/acetate mobile phases for which it was developed. Full details of construction are described in (14).

(iv)   Radioactivity may be monitored with one of a number of commercially available instruments, but for less dedicated use it may be a more economical proposition to collect fractions for counting in an appropriate general purpose scintillation counter. An alternative for some isotopes is a device which allows dried aliquots of effluent to be concentrated and spotted onto paper for drying and autoradiography (15). The film can subsequently be scanned for quantitation.

3.5 Fraction collectors

It is very easy to generate several hundred fractions per day using HPLC. How best to process the resultant tubes deserves careful thought, since subsequent manipulations can easily form an unwelcome bottleneck. A fraction collector which can cope not only with the rapid rate of fraction collection demanded (1 min fractions are usual, 0.2 min not unusual) without losing significant amounts of effluent between tubes but also with a range of different sized tubes and vials is a valuable acquisition. If fractions are to be counted, it saves time to be able to collect directly into scintillation vials, for instance. Because of the large variations in the surface tension of the mobile phase during a gradient-elution drop sizes vary widely. It is therefore more usual to collect by time rather than by drop count.

4. PACKINGS AND COLUMNS

Almost all the column packings available for RP-HPLC are silica based. Some of these silicas are irregular in form, but the majority are spherical, formed by the controlled association of small spherical silica particles to form larger porous spherical aggregates. The mixture of aggregated particles is then fractionated by size to yield the 5, 7 and 10 μm mean diameter packings sold commercially. The mean pore diameter is determined by the diameter of the smaller particles which have aggregated together, which also determines the total surface area contained within each packing particle. All but a few percent of the surface area is contained within the pores. A few supports which are not based on silica have also become available recently: some based on polystyrene which need no additional hydrophobic modification and some based on polyethers suitably modified.

4.1 Reversed-phase packings based on silica

To obtain a reversed-phase packing the surface silica hydroxyls are derivatized with a suitable chlorosilane to introduce the corresponding bonded alkyl or aromatic hydrophobic coating. Originally such reagents as octadecyltrichlorosilane were used, but monochlorosilanes which cannot polymerize nor generate additional hydroxyl groups by hydrolysis are much more common nowadays. For steric reasons not all surface hydroxyls can react with a silane, and the bulkier the hydrophobic group the more difficult it is to achieve maximum coverage. It is usual for manufacturers to follow up reaction with a bulky silane by 'capping' residual accessible silanol groups with a reagent such a trimethylchlorosilane. Besides acting to reverse the polarity of the silica surface the alkyl group serves to protect the silica from aqueous solvents in which it is slightly soluble, even at acid pH, and to mask the unreacted silanols, which are acidic and ionize over a broad range from pH 3.5 upwards, from solutes which may be adsorbed ionically. The bulkier groups seem to be the most effective in the first of these roles, but care should be taken to avoid operating the shorter alkyl supports in purely aqueous eluants for prolonged periods: a few percent of organic solvent improves stability by allowing a protective layer of solvent to be extracted onto the RP surface. Residual surface hydroxyls accessible to solutes can lead to 'mixed mode' adsorption. Where the concentration of silanols is small this behaviour is often characterized by a non-linear adsorption isotherm, that is, an adsorption constant which is concentration dependent. In such circumstances the silanols may act as a non-selective, high affinity, low capacity group of adsorption sites and recoveries of small amounts of protein may be poor, peaks may tail badly or capacity constants may depend on the amount of solute injected. The effect of these silanols can be controlled to some extent either by keeping the pH below 3.5 by adding a small amount of a suitable strong acid such as trifluoroacetic acid (TFA), phosphoric acid, HCl or perchloric acid or much more of a weak acid such as acetic or formic, by raising the ionic strength, or by adding a competing cation such as triethylammonium or ammonium or cetyltrimethylammonium as a 'silanol killer' or even by adding both an acid and a cation. There is evidence that in some cases recoveries of protein are better from supports modified with short alkyl groups than from those carrying long ones, perhaps because of a greater tendency to extensive denaturation. The quality of the parent silica is also an important factor, both for recoveries and efficiency. The importance of pore size has already been mentioned (Section 2.1.2). Commercial products differ quite widely with respect to all these parameters and it is the combination of attributes which determines the effectiveness of a support in actual use. In addition there may be significant variations between batches of a single packing made by a manufacturer. Studies of the effect of different modifying alkyl groups often assume that the contribution of the alkyl groups may be obtained from the comparison of a packing of one type with a packing of another made sometimes by the same, but often by a different manufacturer. Such assumptions do not take account of possible differences in the reagents used for modification, ligand density, capping or the quality of the parent silica. It is also worth remembering that it is the work of a moment to convert even the best of columns into a totally useless one.

4.1.1 *Choice of packing*

Table 1 lists a number of commercially available wide-pore packings, mostly intended for protein use. In addition to RP supports several packings intended for hydrophobic interaction chromatography have been included as they may sometimes be useful used in a RP mode for very hydrophobic proteins. It is not possible to recommend a single brand of packing which is ideal for all circumstances. Some small proteins can be

Table 1. List of reversed-phase column support materials—large pore packings.

Name	d_p (μm)	Mean pore diameter (nm)	UK supplier: comments
Silica based			
C₁			
Astec 300 A C1[a]	5	30	Technicol
Spherisorb S5 (10)X C1[a]	5(10)	30	Phase Sep
Ultropak TSK TMS−250	10	25	LKB
C₃			
Bakerbond wide pore			HPLC Technology:
HI propyl[a]	5	30	really HIC column
Ultrapore RPSC	5	30	Anachem, Beckman
C₄			
Apex WP C4	7	30	Jones Chromatography
Aquapore BU-300	10	30	Anachem, Pierce
Astec 300A C4[a]	5	30	Technicol
Bakerbond Wide Pore			
Butyl[a]	5	30	HPLC Technology
Hipore RP-304	5,10	33	Biorad
Hypersil WP Butyl[a]	5	30	Shandon
Nucleosil 300 C4[a]	5,7,10	30	Camlab
Supelco LC-304	5	30	Supelchem (R.B.Radley and Co.)
Vydac 214 TP[a]	5,10	30	Technicol
C₆			
Spherisorb S5(10)X C6[a]	5(10)	30	Phase Sep
C₈			
Apex WP C8	7	30	Jones Chromatography
Aquapore RP-300	10	30	Pierce
Astec 300A C8[a]	5	30	Technicol
Bakerbond Wide Pore			
Octyl[a]	5	30	HPLC Technology
Hypersil WP Octyl[a]	5	30	Shandon
Lichrospher CH8/II	10	50,100,400	Merck, BDH
Nucleosil 300 C8[a]	5,7,10	30	Camlab
Pro RPC	5	30	Pharmacia: Glass columns
Protesil 300 Octyl	10	30	Whatman
Supelco LC-308	5	30	Supelchem (R.B.Radley and Co.)
Zorbax PEP-RP1			Dupont
C₁₈			
Apex WP C18	7	30	Jones Chromatography
Astec 300A C18	5	30	Technicol
Bakerbond Wide Pore			
Ocadecyl[a]	5	30	HPLC Technology
Hipore RP-318	5,10	33	Biorad
Nucleosil 300 C18[a]	5,7,10	30	Camlab

Table 1. (*continued*)

Name	d_p (µm)	Mean pore diameter (nm)	UK supplier: comments
Spherisorb S5(10)X C18[a]	5(10)	30	Phase Sep
Supelco LC-318	5	30	Supelchem (R.B.Radley)
Vydac 201 TP[a]	5,10	30	Technicol: non-endcapped
Vydac 218 TP[a]	5,10	30	Technicol: end-capped
Phenyl			
Apex WP Phenyl	7	30	Jones Chromatography
Aquapore PH-300	10	30	Pierce
Astec 300A Diphenyl[a]	5	30	Technicol
Bakerbond Wide Pore Diphenyl[a]	5	30	HPLC Technology
Nucleosil 300 C_6H_5[a]	7	30	Camlab
Protesil 300 Diphenyl	10	30	Whatman
Supelco LC-3DP	5	30	Supelchem (R.B.Radley)
Vydac 219 TP[a]	5	30	Technicol
Cyano			
Apex WP CN	7	30	Jones Chromatography
Bakerbond Wide Pore cyanopropyl[a]	5	30	HPLC Technology

Name	Function	d_p (µm)	Mean pore diameter (nm)	UK supplier: comments
Non silica based				
Polyether based				
Biogel TSK RP+	'High density' phenyl	10	100	Biorad
TSK Phenyl 5PW	Phenyl	10	100	Anachem,Biorad: really HIC column
Polystyrene based				
PRLPS-300A	Phenyl	10	30	Polymer Labs

[a]Available as loose packing—rest available as packed columns only.
This list is not intended to be exhaustive and there may well be manufacturers and suppliers other than those listed.

chromatographed perfectly satisfactorily on normal (small-pore) supports when other large-pore supports give inferior results. For larger proteins (mol. wt $>50\,000$) wide-pore packings are essential. Some simple choices are evident. For use at pH >7 one of the non-silica based packings should be chosen. For those who pack their own columns the choice is similarly restricted to those manufacturers who sell their packings loose. If a labile biological activity is to be preserved a short alkyl chain packing is probably to be preferred. The different hydrophobic modifying groups exhibit different selectivities and a separation which is incomplete on one support can be complemented by re-running the unresolved components on another support under the same separation conditions.

Following a C_8 or C_{18} modified support one might choose a phenyl or cyano or C_1 modified one.

4.2 **Column dimensions and capacities**

As mentioned in Section 2.1.3, unless the separation is an isocratic one, the only reason for using a long column for protein separations is to increase capacity. For normal purposes a 15, 10 or even 5 cm column provides adequate resolution and a considerable adsorption capacity. For instance, a 7.5 cm, 4.6 mm i.d. column of a 30 nm pore size packing can adsorb up to 50 mg of cytochrome *c* or lysozyme (16). High loadings have surprisingly little effect on separation so long as a reasonably steep gradient is employed. Since in such conditions the zone of adsorbed protein may extend over a significant proportion of the column length, peak broadening becomes a more serious problem if the elution conditions are close to isocratic. For small scale preparative work it may be cheaper to scale up to a longer column, say 25 cm, though more logical to try a wider i.d. column of the same dimensions.

5. MOBILE PHASES

The mobile phase composition is the most readily changed variable in a RPC separation. In normal usage the mobile phase consists of a mixture of water, a miscible organic solvent and dissolved buffers and salts. Some buffer, such as a low concentration of a strong acid, or salt is essential: if a protein is adsorbed in the absence of salt or acid no increase of proportion of organic solvent will elute it (sometimes a protein will not be adsorbed at all unless conditions are such as to at least partially denature it). In addition some of these or additional components—either solvents or solutes—may be added in order to affect the separation. This may be to delay or accelerate elution, to improve peak shape or to adjust the elution position of some components with respect to others (thus affecting the 'selectivity'). In practice little use is made of small amounts of solvent additives. Mixtures are sometimes used and a change to another organic component may usefully alter selectivity. Other important variables are the pH of the mobile phase, normally but not necessarily in the range $2-4$, and the concentration and type of ions present. The ionic strength is a factor in limiting 'mixed mode' retention involving silica hydroxyls, but the choice of ions can in addition affect the solubility and stability of the protein in the mobile phase and, through ion-pairing, the distribution of the protein between stationary and mobile phases. The net effect depends on the combination of column packing and mobile phase.

5.1 **Detection**

For detection in the UV it is obviously necessary to choose solvents with sufficient transparency. Most proteins have an absorption maximum at about 280 nm which falls to a minimum at 254 nm, a popular wavelength for fixed wavelength HPLC detectors. The highest sensitivities are obtained by monitoring the strong absorption bands which peak below 220 nm. For the hormone parathyrin, for instance, the limits of detectability with conventionally sized columns are about 10 g at 280 nm, 250 ng at 210 nm and about 3 ng employing endogenous tryptophan fluorescence with excitation at 215 nm. During a gradient there is usually an associated change in background absorption. This may be caused directly by the increase in the concentration of organic component or by the effect of this concentration change on other mobile phase components. In addition the refractive index change affects the focal length of the 'liquid lens' set up by flow through the detection cell. Usually not much can be done about the flow-cell design, but careful choice of wavelength and pH can minimize the change in absorption spectrum of mobile phase components. For TFA and acetonitrile the optimum wavelength is 215 nm, for heptafluorobutyric acid (HFBA) it is 219 nm (17). Alternatively the concentrations of other components in the mobile phase can be adjusted to compensate or a small amount of a UV absorbing solute added to one or other reservoir. For example, for a TFA/acetonitrile system at 210 nm the rise in background can be partially counteracted by using 0.13% TFA as the lower limit buffer and 0.1% TFA/80% acetonitrile as the upper.

Water for systems to be monitored in the UV needs to be free of trace UV absorbing organic substances which may be adsorbed on the column during equilibration and the early part of a gradient and eluted as 'ghost' peaks later on. It is sometimes very difficult to avoid such contributions which may come from mobile phase constituents other than water. Water should either be freshly distilled and stored out of contact with plastic or de-ionized and passed through a cartridge especially designed to remove trace organics. Alternatively commercially supplied HPLC grade water can be used, or the aqueous component of the mobile phase pumped through a RP-HPLC column or passed through a bed of coarse RPC adsorbent before use. Such ghost peaks characteristically increase in size with the length of the equilibration time between runs, and have absorption spectra markedly different from that of a protein. Often these problems disappear if the gradient can be started from 15−20% of organic solvent, an eluant strength high enough to avoid concentration of small solutes on the column top and if shorter gradients are used rather than ones covering a large range of solvent strength. Interference problems of this kind are less severe at 280 nm than at 210 nm and do not occur with more specific detection methods such as intrinsic tryptophan fluorescence or post column reaction with fluorescamine. If the column effluent is to be reacted with fluorescamine or *o*-phthalaldehyde/mercaptoethanol (which is much less sensitive than fluorescamine for proteins) then solvents and solutes should be free of amine containing impurities. A common technique is to re-distil from ninhydrin. In general solutes should be of the best quality practicable and kept under appropriate conditions. Our worst problems have been with triethylamine-containing systems and with NaOH.

5.2 Major organic components of the mobile phase

A list of the organic solvents commonly used in the RPC of proteins is given in *Table 2*. The most popular are acetonitrile, propan-1-ol and propan-2-ol. The two chief differences in practical use are (i) that the propanols are much more viscous than acetonitrile, giving higher back pressures—not significant when slow flow-rates and short columns are the norm—and (ii) that the acetonitrile concentration needed to elute a protein is considerably higher than the equivalent propanol concentration, something like half as much again and thus proteins are more likely to be seriously denatured using acetonitrile. There is also evidence that acetonitrile is intrinsically a more powerful denaturant than the alcohol. This may account for the observation that recoveries are

Table 2. List of major organic components of the mobile phase.

Organic solvent	Boiling pt/ freezing pt (°C)	UV[a] cut-off (nm)	Comments[b]
Acetonitrile	82/−42	190	Good UV transparency, but far UV grade essential. Low viscosity. Water mixtures freeze in all proportions and can be lyophilized. Expensive. More denaturing than propanols. Dried residue may be toxic to cells.
Ethanol	78.5/−115	210	Viscous, cheap, rather less effective eluant then acetonitrile. Not popular.
Methanol	64.7/−98.8	205	Half to two-thirds eluting power of acetonitrile. Water mixtures quite viscous. Cheap. Rarely used.
Propan-1-ol (*n*-Propanol)	97.8/−127	210	Very viscous, adequate UV transparency. Good eluting power (> acetonitrile).
Propan-2-ol (*iso*-propanol)	82.3/−89	210	Very viscous, adequate UV transparency. More powerful eluant than acetonitrile, less powerful denaturant. Freezes up to 25% with water
Butan-1-ol (*n*-butanol)		210	Occasionally as part of mixture.
2-methoxy ethanol (methyl cellosolve)		220	Occasionally as part of mixture.

[a]10% transmission (32).
[b]'freezing' refers to behaviour in dry ice/alcohol freezing mixtures.
The order of eluting power is: $CH_3OH < C_2H_5OH < CH_3CN < i\text{-}C_3H_7OH < n\text{-}C_3H_7OH$.
For the alcohols this also corresponds to the order of denaturing power (23).

often better in propanol-containing mobile phases, particularly for the more 'difficult' model proteins such as ovalbumin. A common separation strategy is to follow a separation on a C_8 and C_{18} support in an acetonitrile system by a second separation on a cyano support in a propanol-containing one. All the solvents listed in *Table 2* can be removed in a vacuum oven. Acetonitrile and isopropanol, at the concentrations at which proteins are eluted, can be frozen in dry ice/alcohol freezing mixtures and lyophilized, though it is important to protect the vacuum pump with an efficient cold trap. All except butanol and 2-methoxyethanol are volatile enough to be fairly readily removed in a stream of nitrogen. Solvents should be of the highest quality practicable. Very often small quantities of protein need to be prepared and the effect of trace levels of aldehydes or acrylic acid, for instance, in the organic component can seriously impede the already difficult enough task of recovering activity. In at least one case, the dried residue of an acetonitrile-containing mobile phase was found to be toxic to cells when the equivalent propanol mixtures were non-toxic.

5.3 **Minor components of the mobile phase**

The most popular minor components of the mobile phase are the perfluoroalkanoic acids—particularly TFA and HFBA—at pH 2. These appear to help solubilize proteins in organic solvents. Proteins are eluted later than with equivalent concentrations of phosphoric acid, slightly more retained with HFBA than with TFA, probably due to the effects of ion-pairing. Other popular additives are pyridinium acetate buffers and buffers based on phosphoric acid, particularly triethylammonium phosphate (TEAP). The former are completely volatile but require detection by some technique such as post-column derivatization with fluorescamine (14). TEAP is useful in circumstances where tailing is suspected to be due to non-specific interactions with surface silanols. A selection of systems used is summarized in *Table 3*. Sulphate, phosphate, perchlorate and chloride are all transparent in the far UV, while acetate, formate and fluoroalkanoic acids must be used at concentrations <20 mM if wavelengths below 220 nm are to be used. Pyridine-containing mobile phases cannot be monitored in the UV at all. Ion-pairing agents such as heptane sulphonate, sodium dodecyl sulphate or alkyl ammonium salts may be added to the mobile phase to selectively increase the retention of proteins carrying relatively larger charges of opposite sign. Metal ions such as $CaCl_2$ may be included to enhance the stability of a protein, and low concentrations of non-ionic detergents may improve the behaviour of very hydrophobic proteins such as membrane components. Finally, guanidine hydrochloride or urea or high concentrations (60% by vol.) of powerful solvents such as formic acid may be added to deal with difficult subjects such as virus envelope proteins.

5.4 **Choice of mobile phase system**

The number of different mobile phases used for separating proteins is very large. In most cases conditions have been adjusted to take account of the behaviour of the protein of interest, the characteristics of the column, detector and solvent delivery system in use and, last but not least, the purpose for which the separation is being carried out. Where the purpose of the separation is primarily preparative, the mobile phase is best based on one of the three totally volatile buffer systems listed in *Table 3*. All of these are miscible in all proportions with the organic components and can be removed by lyophilization or drying in a vacuum oven. If it is important to preserve labile biological activity a system which maintains a neutral pH, or starts at neutral pH, using propanol to keep the concentration of the organic component as low as possible should be tried. If these factors are not so important, then systems based on involatile components, of which triethylammonium phosphate is one of the most popular, may give better peak shapes or recoveries. It is important to remember that these salts are not soluble in high concentrations of organic solvents. The exact limit depends on the counter-ion, pH and temperature, but problems of precipitation usually occur over the range 60−80% acetonitrile, for instance. It is safer to make the concentration of organic component in the upper limit solvent less than the level at which precipitation occurs. For this reason it is probably best to put all components into both lower and upper limit solvents and not to have an aqueous buffer solution as one and a pure organic solvent as the other. Precipitates can be extremely damaging to expensive items like pump plungers.

Table 3. List of minor components of the mobile phase.

System	Concentration	Comments
Volatile, low pH:		
Trifluoroacetic acid (TFA)	5−50 mM	0.1% v/v usual. Volatile, miscible in all proportions. Good UV
Heptafluorobutyric acid (HFBA)	10 mM	As TFA. Slightly longer retention times
(Also pentafluoropropionic, pentadecafluorooctanoic)		
Pyridinium formate (acetate)	0.25−0.5 M py ⟶ pH 4−5	UV opaque. Use with post column reaction system. Totally volatile
Triethylammonium formate/acetate	25 mM−0.25 M TEA ⟶ pH 5−6	Poor UV at high concentrations. Usually not as effective as TEAP
Non-volatile, low pH:		
Phosphoric acid	10 mM	Excellent UV. Non-volatile. Lower retention times than TFA. Modest buffering power. Add $(NH_4)_2SO_4$, NaCl or $NaClO_4$ to control silanol related trailing (maximum CH_3CN = 60%)
Triethylammonium phosphate (TEAP)	0.1−0.25 M H_3PO_4 ⟶ pH 2.5−3	Good UV. Excellent silanol killer. Does not necessarily give better separations than TFA. Relatively slow to re-equilibrate. Good quality TEA preferable. Powerful buffering capacity. (Maximum CH_3CN = 70−80%)
0.155 M NaCl/HCl	pH 2.1	Low buffering capacity. NaCl non-toxic, relatively little interference with subsequent procedures. Cl^- corrosive to stainless steel (see Section 3.1.1) (Maximum CH_3CN = 80%)
(Others including 0.1 M NaH_2PO_4/H_3PO_4 pH 2.1, $NaClO_4$ pH 2)		
Neutral:		
TEAP pH 7	25 mM TEA	Good UV. Up to 80% CH_3CN.
Phosphate pH 6−7	0.01−0.1 M	Excellent UV. Limited solubility— maximum 45−50% CH_3CN. Can combine dilute phosphate with neutral salt such as NaCl
Ammonium acetate pH 7 (24)	0.01−0.1 M	Gradient to 15 mM TFA pH 2/CH_3CN. Good recoveries of intact glycoprotein hormones
Difficult molecules:		
HCOOH/*i*-PrOH (25)	60% formic	For viral polypeptides
HCOOH/TEA	6 M/0.13 M pH 1.5	asp−pro cleavage in myoglobin!
TFA/3 M urea (27) or 3 M guanidine HCl	TFA ⟶ pH 3	Successfully eluted otherwise irreversibly bound glycopeptide

Table 3. (*continued*)

System	Concentration	Comments
Neutral salts:	Maximum % of CH₃CN	
Na₂SO₄	60	
(NH₄)₂SO₄	60	These salts are all transparent
NaCl	80	at 210 nm
NaClO₄	70	

For preparative work it is usually necessary to include a de-salting step if a non-volatile system is used, though for operations such as amino acid analysis and sequencing it may be possible to directly analyse a sample run in a dilute phosphoric or phosphate system. For other kinds of post chromatographic analysis, phosphates may possess too great a buffering power and phosphate and sulphate are both very toxic to cells. If examination of the effects on cell culture or a radioimmunoassay are planned, a system of low buffering power containing a relatively innocuous neutral salt like NaCl adjusted in pH with HCl is more appropriate if a suitable volatile system cannot be used. At pH <4 phosphate containing systems lyophilize to a syrup, while neutral salt systems give a solid.

The best general advice is to try a TFA/acetonitrile or propanol system first and to modify or experiment with others on the basis of this experience. It may be necessary to increase the concentration of TFA from the normal 10 mM (0.1% v/v)—this depends on the protein and on the column packing used.

6. ACHIEVING AND MODIFYING A SEPARATION

6.1 Sample preparation

One of the charms of RPC is the undemanding nature of the requirements surrounding the sample to be injected. Large volumes of dilute aqueous solutions of a protein may be concentrated on the top of a RP column simply by pumping on (see Section 3). For small volumes (up to 50 μl) it often makes little difference whether the sample is at the pH of the mobile phase or not. Neither high salt concentrations nor high urea or guanidine concentrations interfere with sample application, though they may affect the elution behaviour of a protein, particularly if it has been dissolved in a denaturing solvent for some time. On the other hand, it is important that a protein should be rapidly bound by the support and that precipitation should not take place. The only components which have a serious effect on application are organic solvents in high concentrations or detergents in large amounts. For complex samples, and where there is a chance that unwanted components may be strongly or irreversibly adsorbed to the HPLC column, the sample can be pre-fractionated on a small cartridge or column of RP silica or other suitable packing (even simply using loose adsorbent). The suitably clarified sample, preferably acidified, is passed through the RP adsorbent, twice if necessary, the adsorbent is washed and the adsorbed material eluted batchwise with increasing concentrations of organic solvent. The eluant may be based on any of the systems in *Table 3*, but a system containing 1% TFA or other volatile components has obvious advantages. Following elution the fraction may be lyophilized and reconstituted, but better recoveries are achieved if it is simply diluted appropriately or if the organic solvent concentration is reduced by a stream of nitrogen and the fraction then simply loaded onto the analytical column.

6.2 Initial conditions

As with all column chromatography, the first step is to establish conditions for adequate recovery of the components of interest from the adsorbent. This may not always be feasible to establish, but batch elution experiments with commercially available sample preparation cartridges (available from a number of manufacturers, some containing wide-pore RP silica) or small hand packed columns of RP silica can provide valuable pointers to likely conditions. It is essential to include blank gradient runs at regular intervals, firstly to check for ghost peaks due to solvent impurities and secondly to identify components which are being incompletely eluted during one gradient and are reappearing in subsequent chromatograms. Because of evidence that protein recovery is particularly affected by exposure to RP surfaces in the absence of organic modifier it is advisable to ensure that the initial concentration of organic modifier is significant. Few proteins are eluted below 15% acetonitrile or 10% propanol, which therefore represent sensible lower limits. For initial purposes, a linear gradient of 1−3% organic component/min is suitable at a flow-rate of 1−1.5 ml/min for a conventional (4.6 mm i.d.) column. This means that the range of 20−60% acetonitrile or 10−50% propanol, which encompasses the range of solvent strengths over which proteins are usually eluted, can be covered in 20−30 min. The gradient limits can then be adjusted to achieve best results. Often unwanted material eluted late in the chromatogram can be telescoped by making the last part of the gradient steeper than the main part.

6.3 Improving separation

Once the components of interest have been eluted it may still not be possible to obtain a satisfactory separation by simply adjusting the flow-rate and gradient parameters.

6.3.1 *Poor peak shape*

Peak tailing is a common problem. If the system is a TFA one, then increasing the TFA concentration from 10 mM to 20 mM or 50 mM may improve peak shape to an acceptable degree and thus the resolution between components. The disadvantage is that the UV monitor wavelength may have to be increased with the increase in TFA concentration. Tailing of this kind is often associated with high levels of accessible silanols on the silica surface. Switching to a fresh column (if using an old one) or to another make or type of column packing may cure the tailing. For diagnostic purposes, the addition of 50 mM ammonium sulphate or sodium perchlorate to the mobile phase may lead to a dramatic improvement in performance, indicating almost certainly that silanol interactions are the cause. Proteins differ markedly in the degree to which they are affected by such interactions and it may be necessary to use a mobile phase containing an effective competing ion such as triethylamine, or to increase the ionic strength or lower the pH. The non silica-based columns listed in *Table 1* should not show this particular type of behaviour.

Other causes of poor peak shape may be associated with slow transitions between protein forms which bind to the packing with different affinity constants. In the simplest case this may be either because the protein has bound in an unfolded form and has re-folded during elution (9)—in which case fronting is the most likely outcome—or because the protein has bound in a less unfolded form and is unfolding further during elution, when trailing may be observed. The solution, if such a simple analysis can be applied, is to try to shift the equilibrium decisively one way or the other. Lowering pH, raising the temperature and switching to an organic component like acetonitrile which can be used in higher concentrations usually shift the equilibrium in favour of unfolded protein and increase the rate of unfolding. Increasing the flow-rate or steepness of the gradient may shorten the time available for such transformations if they are occurring in the liquid phase. Trailing peaks are also typical of large proteins chromatographed on small-pore packings. Finally the classical cause of tailing in the separation of small solutes is a void at the column top: the solution is to remove the inlet fitting and top-up the column material with a methanol slurry of replacement RP material or fine glass beads. If necessary badly discoloured packing can be removed at this stage.

6.3.2 *Poor recoveries*

Poor recoveries, especially of small amounts of material ($< 20 \mu g$), may again be due to interactions with silanols and the cures are the same as in Section 6.3.1. Often, however, the reasons are to do with the mechanism of adsorption and can be quite complex. Recoveries may often be improved by a switch from acetonitrile to propanol or *vice versa*, or by a switch to or from a TFA-based mobile phase. Proteins may undergo slow changes while adsorbed to the surface, which may occur more readily if the mobile phase has no organic component. Such losses will increase with the time the protein is adsorbed and be reduced if the gradient is made steeper or if the starting conditions are altered to include significant concentrations of organic component. Slow conformation changes which may affect losses occur in denaturing solutions such as guanidine before ever the protein is subjected to chromatography. In such cases the protein should be dissolved as soon as possible before injection. Other changes may occur only in the liquid phase and thus the time between the development of significant concentrations of protein in the liquid phase and the time the protein emerges from the column—determined by the steepness of the gradient and the flow-rate—should be minimized. Finally, lower losses may occur on silicas modified with shorter alkyl chains such as C_4, C_3 or C_1. It is not uncommon to find that proteins recovered in poor yield reappear in subsequent blank gradient runs. A heavy load of ovalbumin (21 mg) continued to appear for at least 11 subsequent gradients (18).

6.3.3 *Altering resolution*

If reducing the flow-rate and the slope of the gradient fail to deliver adequate resolution then some other change to the system must be made. The simplest, and most expensive, is to change the column. If using a C_8 packing try a phenyl, a cyano or a C_1. Alternatively change the organic component, for example from acetonitrile to propanol. The third simple variable to alter is the pH, especially if using a buffer such as phosphate. Addition of a counter-ion such as TEA may make a considerable difference to the order in which components are eluted and this may be enhanced with very hydrophobic ions such as cetrimide, HBFA, HDFOA or even SDS. Finally an increase in ionic strength may sharpen up peaks dramatically.

6.4 **Other problems**

6.4.1 *Maintaining activity*

Some proteins withstand the conditions of RPC with no apparent loss of biological activity. Many growth factors, for instance interferons and interleukins, peptide hormones such as insulin and even ribosomal proteins can be chromatographed in a comparatively cavalier manner without loss of activity or ability to re-associate. Many proteins retain antigenic activity without special care being taken, but others, especially hydrophobic proteins which are difficult to elute from RP columns, may emerge no longer antigenically recognizable. Enzymic activity is much more labile, though some enzymes may be recovered in high yield. In general the best approach seems to be to try to use conditions which promote the integrity of the folded protein structure, such as keeping the pH as high as possible, the temperature as low as possible, adding stabilizing metal ions or co-factors to the mobile phase (such as Ca^{2+} in the case of trypsin) and trying to ensure that the eluate is collected into an environment which encourages re-folding as rapidly as possible. The eluate may thus be collected into a suitable buffer containing metal ions, substrates or inhibitors, co-factors (which may easily be lost during chromatography) and such structure stabilizing components as glycerol, and kept at an appropriate temperature.

6.4.2 *Difficult proteins*

Some very hydrophobic proteins may be very difficult to elute from a RPC column at all. Certain viral proteins and membrane proteins are examples. If no suitable technique other than RPC presents itself, then one of the last three systems listed under the heading of 'Difficult molecules' in *Table 3* may succeed where less powerful eluants fail. These are more difficult to monitor at high sensitivity by UV absorption. 60% formic acid is a strong enough acid to run the risk of serious de-amidation or cleavage at acid labile ...asp−pro... sequences. Nevertheless these three systems are based on powerful proteins solvents—either strong formic acid or urea or guanidine containing mixtures. An alternative is to use a column of lower hydrophobicity such as one of the HIC columns listed in *Table 1* in combination with a conventional solvent system such as TFA−acetonitrile. This combination has been reported to give better recoveries of hydrophobic proteins than is possible on conventional RP columns (19). Finally the inclusion of detergents such as Triton X100, BRIJ 35 or Octyl glucose (20,21) may help to reduce the retention and improve the resolution of hydrophobic proteins on RPC columns.

7. APRES CHROMATOGRAPHY

7.1 Scale-up

One of the great advantages of HPLC in general is the ease with which analytical runs of a few μg can be scaled-up to preparative runs of a few mg. For larger quantities— less than 100 mg—it is simplest to repeatedly inject and collect the portion of interest, particularly if this can be carried out with the help of an automatic injector and an intelligent fraction collector with peak detection. For quantities greater than 100 mg it is better to change to a larger column. This should ideally be of the same packing and particle size as was used analytically, if budget permits, otherwise as small a particle diameter as possible. It may also be worth switching to a cheaper small-pore packing, particularly for smaller proteins. For very large-scale work, initial fractionation can be carried out with cheap preparative packings such as Lichroprep C_8 or C_{18}, $5-25$ μm (Merck), which can be gravity packed into glass columns and operated at low pressure.

7.2 Dealing with fractions

For some analytical purposes it is necessary to collect fractions and examine them subsequently, often by immunoassay or bioassay. Often the amounts of material involved are very small and it is important to ensure that active protein is not adsorbed to the wall of the tube into which it is collected. Plastic tubes may be preferable to glass for this reason and fractions may be collected into an inert bulking agent such as RIA grade BSA (50 μl of 1 mg/ml) to reduce non-specific adsorption. One percent TFA is a useful solvent for recovering small quantities of material from glass tubes. Organic solvents interfere with binding in immunoassays and must be completely removed for high sensitivity work. Often the amounts of material available are large compared with the sensitivity of the assay, and the solvent can simply be diluted out, or a standard curve obtained in the presence of an appropriate concentration of solvent. Generally it is better to avoid strongly buffering involatile mobile phase components like phosphate. The great speed at which an HPLC system can produce fractions for further processing has been mentioned above (Section 3.1.4). In planning analytical procedures the rapidity with which tubes can be processed is an important consideration.

7.3 De-salting

If it is not possible to use a totally volatile mobile phase it may be necessary to de-salt a preparation following HPLC. This can be carried out simply by traditional methods such as dialysis or solvent exchange on a small column of one of the traditional gel permeation materials. However, the sample may also be diluted, adsorbed onto a small column of RP material and eluted by a rapid gradient of propanol in 0.1% TFA, or by stepwise elution from a suitable RP sample clean-up cartridge. With microbore columns it is possible to obtain the sample in a few μl by this method (16).

7.4 Losses of material

The most likely cause of protein loss is the packing inside the column, a subject dealt with in Section 6.3.2. Proteins may also bind to or precipitate on stainless steel surfaces. The column walls can contribute very little to losses both because surface area is low and because only a small proportion of solute molecules ever make contact with the walls. Frits have considerable surface area and losses due to adsorption or denaturation on the column frit may be significant. Meshes have much lower surface areas and present a lower resistance to flow. Under some conditions significant losses of proteins on the walls of the stainless steel capillary tubing used to interconnect HPLC equipment may occur (22). Losses in both these ways may be exacerbated if the protein is displaying a tendency to precipitate—for instance if the pH of the injected solution and the mobile phase are on opposite sides of the protein's isoelectric point. For moderate pressures porous PTFE frits may be used and may prove preferable to stainless steel. On the other hand, the PTFE tubing often used on the low pressure side of an HPLC column may also adsorb proteins, at least iodinated ones. These problems are not encountered often, perhaps because not recognized, but it is as well to be aware of them.

7.5 Modification of protein during chromatography

The most likely changes which may occur to a protein are denaturation and aggregation or, more likely, dissociation. However, other changes can occur during RPC. The acidic mobile phases commonly used may encourage de-amidation or even, in some cases, cleavage at acid-sensitive sequences such as asp−pro. Strongly acidic alcoholic solutions may esterify carboxyl groups. A common modification during handling is the oxidation of methionine residues to methionine sulphoxide. Since methionine is relatively hydrophobic this is more common in peptides than proteins, where methionines are more likely to be buried out of harm's way. All of these may lead to a change in retention. The classical and essential test of any suspected irreversible change is to re-inject the collected peak to confirm that it re-emerges in the same place.

7.6 Reproducibility

For isocratic systems it is very important to make up mobile phases accurately. It is best to dispense solvents by weight rather than by volume. Evaporation of the organic component may have a significant effect on retention, especially if solvents are over-enthusiastically de-gassed by sparging with helium. Changes in ambient temperature may also affect retentions. In gradient separations it is important to allow sufficient time for re-equilibration between runs—items such as mixing chambers contribute a significant volume which may introduce a considerable lag in reaching equilibrium conditions. This is usually obvious from baseline changes if the UV absorbance is monitored continuously between runs, but may be underestimated if monitoring only occurs during the gradient, as is common if an integrator is used. It may be best to include a standard re-equilibration segment at the end of every gradient profile to enforce good habits.

7.7 Temperature

An increase in temperature usually decreases retention in RPC. However, since temperature changes may also affect protein structure the effect on RPC behaviour of a protein is less predictable—it may, for instance, lead to an increase in retention. Such changes may be manipulated to improve separation (see Section 8.1) or may be a source of irreproducibility which may be eliminated by thermostatting column and injector.

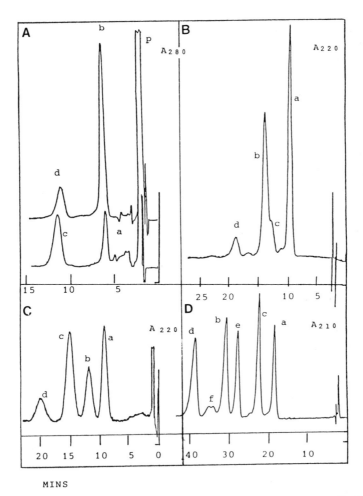

MINS

Figure 2. Isocratic reversed-phase separation of insulin mixtures in four different systems. Conditions: **Panel A**: Column, 10 × 0.5 cm SAS−Hypersil (Cl, 5 µm). Mobile phase, 1% w/v cetrimide in 0.1 M Tris−HCl 10 mM EDTA pH 7.5:methanol, 27:73 (28); temperature, ambient; sample, **upper trace**, bovine insulin soluble formulation; **lower**, porcine insulin neutral formulation following accelerated degradation. **Panel B**: 15 × 0.46 cm ODS Hypersil. Mobile phase, 5 mM tartaric acid−0.1 M ammonium sulphate pH 3: acetonitrile, 73:27 (29); temperature, ambient; sample, international reference preparation of insulin for bioassay (established 1956). **Panel C**: 15 × 0.46 cm ODS Hypersil. Mobile phase, 5 mM tartaric acid−0.1 M ammonium sulphate pH 3: acetonitrile 75:25 containing 14 µM cetrimide (30); temperature, ambient; sample, artificial mixture of bovine and porcine insulins and monodesamido insulins. **Panel D**: 25 × 0.46 cm Ultrasphere ODS. Mobile phase, 0.1 M dihydrogen phosphate/phosphoric acid pH 2: acetonitrile 70:30; temperature, 45°C (30); sample, artificial mixture. Peak identities: **a**, bovine native; **b**, porcine native; **c** bovine [21]A monodesamido; **d**, porcine [21]A monodesamido; **e**, human native; **f** human [21]A monodesamido; **p**, preservative.

8. EXAMPLES

The following examples, mostly of fairly small proteins, have been chosen to illustrate some of the points made in the main text.

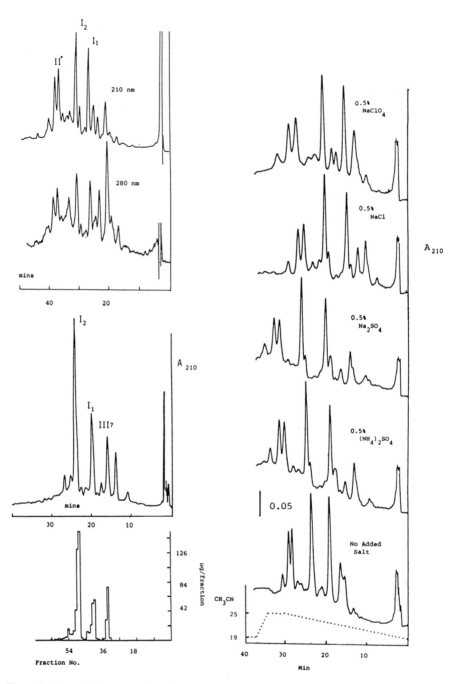

Figure 3. Reversed-phase separation of a mixture of partially purified porcine neurophysins. **Panel (a)**, identification of neurophysin components by 280/210 nm ratios. Sample, porcine neurophysin I and II; column, 15 × 0.2 cm Nucleosil 5 μC_8; flow, 0.2 ml/min; mobile phase, A = 0.13% TFA, B = 80% CH_3CN, 0.1% TFA; gradient, linear, 38% – 50% B over 40 min. **Upper trace**, 20 μg sample, 210 nm, 0.512 AU FSD. **Lower trace**, 50 μg sample, 280 nm, 0.032 AU FSD. **Panel (b)**, identification of components by

radioimmunoassay. Sample, neurophysin I + III mixture; column, 10 × 0.5 cm Nucleosil 5 μC_8; flow, 1.0 ml/min. A = 0.1 M NaH_2PO_4 pH 2.1, B = 60% CH_3CN/40% A. Gradient, linear, 40−50% B over 30 min. **Upper trace**, 210 nm, 0.5 FSD. **Lower trace**, immunoassay versus anti-porcine neurophysin antibody (μg equivalents). **Panel (c)**, effect of salt on separation. Sample, neurophysin I and II [same sample as (**a**)]; column, Aquapore RP-300, 25 × 0.46 cm; flow, 1.5 ml/min. All traces, 215 nm, 0.05 FSD. All mobile phases contained 0.05% H_3PO_4 with addition of 0.5% w/v of the salts shown except for the lowest trace. Gradients were of the same shape as shown for the lowest trace, except that limits were as follows:

Salt added	Initial %CH_3CN	Final %CH_3CN
None	19	25
$(NH_4)_2SO_4$	21.5	27.5
Na_2SO_4	21.5	27.5
NaCl	23	29
$NaClO_4$	29.5	35.5

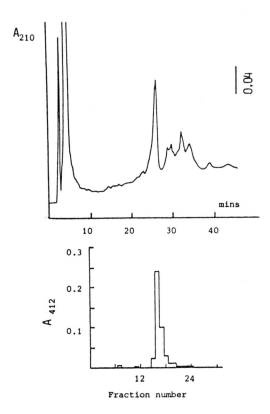

Figure 4. Reversed-phase separation of trypsin in pancreatin. Column, 15 × 0.2 cm Hypersil WP Octyl. Conditions, mobile phase: A = 0.13% TFA, B = 80% CH_3CN, 0.1% TFA; gradient, linear, 30−75% B in 40 min (24−60% CH_3CN), flow, 0.2 ml/min; sample, extract of 200 μg pancreatin in 100 μl 10 mM HCl−mM $CaCl_2$. One min fractions were collected over the period shown, frozen, lyophilized and reconstituted in 0.1 ml mM HCl−mM $CaCl_2$. 20 μl aliquots were mixed in a titre plate well with 200 μl 1 mg/ml benzoyl arginine *p*-nitroanilide. Following incubation at 37°C for 30 min the plate was read at 412 nm in a plate reader.

153

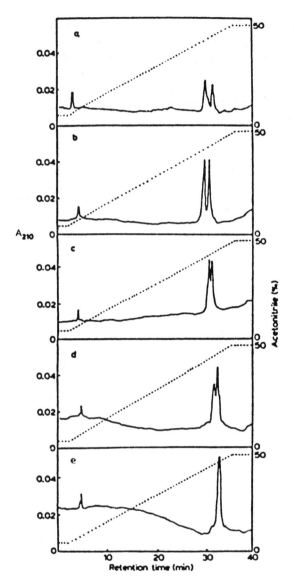

Figure 5. Effect of pH on the reversed-phase chromatography of thyrotropin. Column, 25 × 0.46 cm Aquapore RP-300; mobile phase, A = 0.025 M orthophosphoric acid adjusted with sodium hydroxide to pH 2.0 (**a**), 3.0 (**b**), 4.7 (**c**), 6.0 (**d**), 7.0 (**e**), B = acetonitrile. The linear gradient from 5–50% acetonitrile is indicated by the broken line. Reprinted from (31) with the permission of the publishers and authors.

8.1 Insulin

Figure 2 (p. 151) is firstly a demonstration that small proteins can be usefully chromato-graphed isocratically and secondly a demonstration of how separations may be manipulated by the addition of ion-pairing agents—in this case the very hydrophobic quaternary ammonium ion cetrimide (cetyl trimethylammonium bromide). The order of elution of monodesamido bovine insulin and porcine insulin (labelled c and b, respectively) is reversed by the addition of low concentrations of cetrimide (panels B and C) and the resolution of bovine and porcine insulin almost completely abolished by a high concentration (panel A). In fact the separation in panel A looks very similar to separations of insulin mixtures by high-performance anion-exchange chromatography. At neutral pH it was essential to include EDTA in the mobile phase to obtain acceptable peak shapes. This was presumably due to the reduction in the availability of Zn^{2+} to participate in the formation of the zinc complex. Panel D shows a separation in a different system of three insulins and their monodesamido forms. In order to achieve a full separation it was necessary to raise the temperature to 45°C. Interestingly enough the retentions of all components increased with temperature up to 45°C, suggesting that the equilibrium conformation was shifting towards a more unfolded form. All these separations were carried out on conventional small-pore column packings. For isocratic work mobile phases are best made up by weight rather than by volume.

8.2 Porcine neurophysins

The neurophysins are synthesized in the posterior pituitary as part of the same precursor sequences as oxytocin and vasopressin, with which they are associated within the neurohypophysis. Porcine neurophysins I and III bind vasopressin, neurophysin II oxytocin. They have relative molecular masses of about 10 000 and are very soluble in organic solvents. The sequence represented by neurophysins I and III is found in three forms: neurophysin I_1, with C-terminal sequence −ser−phe, I_2 with sequence −ser−phe−leu and III which has an additional arg_2ala at the C-terminus. *Figure 3* (p. 152) shows a separation of a partially purified mixture of porcine neurophysins. The preparation used contained components other than neurophysin. The neurophysin-related components could be checked with antibodies raised to neurophysins (panel b) and since the neurophysins contain little or no tryptophan, the neurophysin peaks could easily be distinguished by their 280/210 nm absorption ratios. Here this has been determined by monitoring a duplicate run at 280 nm and higher sensitivity (panel a). Panel c is a demonstration of how the addition of different salts may be used to modify a separation—the effects of varying the salt are most evident in the pattern of minor components. Perchlorate has a particularly dramatic effect on retention (note the gradient limits) which agree with its well-known properties as an ion-pairing agent.

8.3 Trypsin in pancreatin

Pancreatin is a commercial product containing significant amounts of the digestive enzymes secreted by the pancreas. It is used as a food supplement to complement inadequate pancreatic function. *Figure 4* (p. 153) shows a chromatogram of a pancreatin extract, together with the results of assaying the fractions collected with a specific colorimetric substrate for trypsin-like enzymes.

8.4 Thyrotropin (TSH)

Thyrotropin is a member of the class of glycoprotein hormones. These are non-covalently associated dimers consisting of a common α subunit and differing in the structure of the β subunit. Each subunit has a molecular weight of about 15 000 and contains

155

carbohydrate. Under physiological conditions circulating concentrations (10^{-10} M) are well below the equilibrium affinity constant (in the region of 10^{-6} M) and the association is maintained kinetically because of the very slow dissociation rate constant (7). The rate of dissociation is accelerated at acid pH or by denaturing agents. Most published separations have dealt with separations of the subunits. The set of chromatograms illustrated in *Figure 5* (p. 154) show that by varying the pH it is possible to recover the hormone either intact or as the separated α and β subunits.

9. REFERENCES

1. Snyder,L.R. and Stadalius,M.A. (1986) *High-Performance Liquid Chromatography*, **4**, 195.
2. Hearn,M.T.W. (1984) In *Methods in Enzymology*. Jacoby,W.B. (ed.), Academic Press Inc., New York, vol. 104, p.190.
3. Tanford,C., (1961) *Physical Chemistry of Macromolecules*. John Wiley & Sons Inc., London and New York.
4. Knox,J.H. (1977) *J. Chromatogr. Sci.*, **15**, 352.
5. Lewis,R.V. and Stern,A.S. (1984) In *Handbook for the Separation of Amino Acids, Peptides and Proteins*. Hancock,W.S. (ed.), CRC Press Inc., Boca Raton, FL, p.313.
6. Creighton,T.E., (1984) *Proteins*. W.H. Freeman & Co., New York.
7. Parsons,T.F., Strickland,T.W. and Pierce,J.G. (1985) In *Methods in Enzymology*. Birnbaumer,L. and O'Malley,B.W. (eds), Academic Press Inc., New York and London, vol. 19, p. 736.
8. Katzenstein,G.E., Vrona,S.E., Wechsler,R.J., Steadman,B.L., Lewis,R.V. and Middaugh,C.R. (1986) *Proc. Natl. Acad. Sci. USA*, **83**, 4268.
9. Lu,X.M., Benedek,K. and Karger,B.L. (1986) *J. Chromatogr.*, **359**, 19.
10. Ferris,R.J., Cowgill,C.A. and Traut,R.R. (1984) *Biochemistry*, **23**, 3434.
11. Benedek,K., Dong,S. and Karger,B.L. (1984) *J. Chromatogr.*, **317**, 227.
12. Grafl,R., Lang,K., Wrba,A. and Schmidt,F.X. (1986) *J. Mol. Biol.*, **191**, 281.
13. Melander,W.R., Jacobsen,J. and Horvath,C. (1982) *J. Chromatogr.*, **234**, 269.
14. Stein,S. and Moschera,J. (1981) In *Methods in Enzymology*. Pestra,S. (ed.), Academic Press Inc., New York, vol. 79, p. 7.
15. Karmen,A., Malikin,G. and Lam,S. (1984) *J. Chromatogr.*, **302**, 31.
16. Nice,E.C., Lloyd,C.J. and Burgess,A.W. (1984) *J. Chromatogr.*, **296**, 153.
17. Winkler,G., Briza,P. and Kunz,C. (1986) *J. Chromatogr.*, **361**, 191.
18. Pearson,J.D., Lin,N.T. and Regnier,F.E. (1982) *Analyt. Biochem.*, **124**, 217.
19. Goheen,S.C. and Chow,T.M. (1986) *J. Chromatogr.*, **359**, 297.
20. Hearn,M.T.W. and Grego,B. (1984) *J. Chromatogr.*, **296**, 309.
21. Josic,Dj., Reutter,W. and Molnar,I. (1982) In *Practical Aspects of Modern HPLC*. Molnar,I. (ed.), Walter de Gruyter, Berlin, p. 109.
22. Trumbore,C.N., Tremblay,R.D., Penrose,J.T., Mercer,M. and Kelleher,F.M. (1983) *J. Chromatogr.*, **280**, 43.
23. Friesen,H.-J. (1982) In *Practical Aspects of Modern HPLC*. Molnar,I. (ed.), Walter de Gruyter, Berlin, p. 77.
24. Grego,B. and Hearn,M.T.W. (1984) *J. Chromatogr.*, **336**, 25.
25. Heukshoven,J. and Dernick,R. (1982) *J. Chromatogr.*, **252**, 241.
26. Tarr,G.E. and Crabb,J.W. (1983) *Analyt. Biochem.*, **131**, 99.
27. Sharifi,B.G., Bascom,C.C., Khurana,V.K. and Johnson,T.C. (1985) *J. Chromatogr.*, **324**, 173.
28. Corran,P.H., Calam,D.H. (1979) In *Proceedings of the Ninth International Symposium on Chromatography and Electrophoresis*. Frigerio,A. (ed.), Elsevier, Amsterdam, p. 341.
29. Terabe,S., Konaka,K. and Inouye,K. (1979) *J. Chromatogr.*, **172**, 163.
30. Lloyd,L.F. and Corran,P.H. (1982) *J. Chromatogr.*, **240**, 445.
31. Bristow,A.F., Wilson,C. and Sutcliffe,N. (1983) *J. Chromatogr.*, **270**, 285.
32. Bristow,P.A. (1976) *LC in practice* hetp, Macclesfield.

CHAPTER 6

High-performance affinity chromatography (HPAC)

YANNIS D.CLONIS

1. INTRODUCTION

High-performance affinity chromatography (HPAC) combines the inherent speed and resolving power of high-performance liquid chromatography (HPLC) with the exquisite biological specificity of affinity chromatography. This technology has recently attracted considerable interest from research scientists as well as medical and industrial companies in view of its rapid diagnostic potential and capacity of resolving multigram quantities of complex protein mixtures in a short time (1).

Rapid chromatographic separations require that the mobile phase carrying the solutes must move with high flow-rates. Therefore it is usually forced through appropriate microparticulate columns under pressure. Until recently only silica-based supports were commercially available; however, in the last few years a number of purpose-designed synthetic hydrophilic gels have been launched in the market. This chapter will focus attention upon the properties, synthesis, packing and applications of HPAC adsorbents.

2. PROPERTIES AND PRACTICAL CONSIDERATIONS OF COLUMN SUPPORT MATERIALS OR MATRICES USED IN HPAC

The operation of an HPAC separation requires a column support material, termed the matrix, that:

(i) is non-compressible, so that it can be operated under high flow-rates of the mobile phase which usually require high pressures;

(ii) consists of small spherical rigid particles (typically between 5 and 30 μm particle size) with narrow size distribution so that the column exhibits good plate number and low band broadening;

(iii) is hydrophilic but not charged in order to minimize non-specific adsorption;

(iv) is chemically stable during ligand immobilization and hygiene maintainance;

(v) is resistant to chemical and biological degradation;

(vi) possesses wide pores so that the biomolecules can penetrate freely in the interior cavities of the particles without steric hindrance by the matrix itself.

The matrices commercially available fulfil these requirements to different extents; *Table 1* summarizes some characteristics of beaded support materials.

Table 1. Characteristics of some beaded gels suitable for use in HPAC after appropriate derivatization.

Gel	Particle size (µm)	Measure of pore size	Stability pH	max. pressure (bar)	Surface groups	Manufacturer
Silica						
Hypersil WP 300	5 and 10	30[a1]	2–8	1035	Silanol	Shandon (GB)
LiChrospher range	10	30–400[a2]	2–8	–	Silanol	Merck (FRG)
SelectiSpher-10	10	50[a1]	2–8	80	Tresyl-diol-bonded	Pierce (USA)
Ultraffinity-EP	–	–	2–8	–	Epoxy-bonded	Beckman (USA)
Synthetic						
TSK-PW range	from 10 to 20	$(10^3 - 3 \times 10^7)$[b]	1–13	130	Hydroxyl	Toyo Soda (Japan)
Separon H 1000	10	$(<1 \times 10^6)$[c]	1–13	50	Hydroxyl	LIW (Czechoslovakia)
Dynospheres XP-2507	20	5–200[a1]	1–13	140	Hydroxyl	Dyno Particles (Norway)
Eupergit C30 N	30	$(10^3 - 25 \times 10^3)$[c]	1–12	100	Epoxy	Röhm Pharma (FRG)
Natural						
Crosslinked agarose (12%)	3–10	(450×10^3)[c]	2–14	60	Hydroxyl	ref. 7
Crosslinked agarose (12%)	5–40	(450×10^3)[c]	2–14	60	Hydroxyl	ref. 7
Superose 6	13	$(5 \times 10^3 - 5 \times 10^6)$[c]	1–14	15	Hydroxyl	Pharmacia (Sweden)
Superose 12	10	$(<3 \times 10^5)$[c]	1–14	30	Hydroxyl	Pharmacia (Sweden)

[a1]Pore diameter (in nm); [a2]Varies according to gel's grade.
[b]Molecular exclusion limits measured with polyethylene glycol and/or dextran; varies according to gel's grade.
[c]Molecular exclusion limits for proteins.

2.1 Silica matrices

Although a number of inorganic materials capable of forming gels such as silica, titania, zirconia and alumina are potential candidates for the column support gel, silica has been by far the most popular medium. Silica gel is usually produced by precipitating sodium silicate solution with acid, followed by washing the gel and drying it (2). Milling, sieving, agglomeration, etc. then give the various types of commercially available silica gel. Also, finely divided silica gel can be obtained by burning silicon tetrachloride in a hydrogen flame. The surface chemistry of silica is characterized by the presence of siloxane bridges and silanol groups; the latter occur both as free silanol functions and hydrogen-bridged ones at a concentration typically of about $4.8/nm^2$. In equilibrium with the atmosphere the silica gel is associated with about $5-15\%$ of adsorbed/absorbed water. The physically absorbed water can be removed by heating at $110-150°C$ under reduced pressure, whereas, the chemically bonded water requires temperatures higher than $200°C$, a process which leads to the formation of one siloxane bridge for every molecule of water eliminated from two silanol groups. During the manufacturing process of beaded macroporous silica gel, high temperatures are employed; thus, it is advisable first to boil the silica gel in $1-10$ mM HCl prior to other chemical manipulations. In this way most siloxane bridges are converted to silanol groups. This step is not necessary for old silica stocks. The negatively charged silanol groups on the silica's surface, however, endow the gel with considerable non-specific adsorption characteristics. Such deleterious effects are usually circumvented, to some extent, by surface modification (coating) with organosilanes. During the coating reaction with various trimethoxy-organosilanes, part of the surface silanols are derivatized and an almost neutral hydrophilic surface is created possessing appropriate groups for ligand attachment. Nevertheless, during the coating reaction with trimethoxy-organosilanes under aqueous conditions, new silanols are introduced on the 'coated' surface due to partial hydrolysis of the reagent, thus, leading to multi-layer formation. This results in bonded-phase layer thicknesses of up to 2 nm (3). To this end, attempts to cap the remaining free silanol groups after silanization by employing trimethylchlorosilane compromise the hydrophilicity of the silica surface. To avoid silane hydrolysis silica can be effectively bonded with appropriate silanes under dry conditions (4).

The main drawback of silica is its high solubility at pH 8.0 and above, although this problem may partly be circumvented by pre-treatment with mixtures of zirconium salts (5). In addition to this, the availability of beads of porous silica gels possessing only moderate pore sizes (<50 nm; column 3 of *Table 1*) has hindered the development of HPAC into a major purification technique for large biological molecules such as proteins and enzymes. It has only recently been possible to obtain beaded silica gels of wide pore size (>50 nm).

2.2 Synthetic matrices

2.2.1 *Eupergit® C 30N*

This acrylic beaded material contains reactive epoxy groups via which affinity ligands possessing primary amino, hydroxyl or sulphydryl groups can be directly immobilized. Also, the epoxy groups can easily be converted to the corresponding diol (in 1 M NaOH) via which a range of ligands can then be attached. Eupergit C is obtained by co-polymerization of methacrylamide, methylen-*bis*-methacrylamide, glycidylmethacrylate and/or allyl-glycidylether. Due to the nature of monomers, the gel shows an electroneutral and mostly hydrophilic character but some hydrophobic interactions may be expected due to the methyl groups along the polymer back-bone. Some physical properties of Eupergit C are listed in *Table 1*.

2.2.2 *TSK® PW-type*

This vinyl-based beaded gel has primarily been developed for size-exclusion chromatography but its derivatized forms find application in ion-exchange, hydrophobic and affinity chromatography. The gel displays low non-specific adsorption of proteins compared with other synthetic suports and its surface possesses functional hydroxyl groups which can be used to immobilize a wide range of affinity ligands by employing appropriate methods (6). Some properties of the TSK PW gel are shown in *Table 1*.

2.2.3 *Separon® H1000*

Separon gels are prepared by co-polymerization of hydroxyalkylmethacrylate with alkene dimethacrylate. This leads to heavily cross-linked microparticles which aggregate to give macroporous beads with functional hydroxyl groups. *Table 1* gives some properties of Separon H1000.

2.2.4 *Dynospheres® (XP-3507)*

These acrylate-co-polymer beads are macroporous, hydrophilic and possess functional hydroxyl groups for ligand attachment. The main feature of this gel is its extremely low back-pressure due to the unique monosized character of the perfectly spherical particles. Some properties of this gel are shown in *Table 1*.

2.3 Natural matrices

Although such matrices have not been used for HPAC so far, one should note that cross-linked agarose gel of high agarose content (e.g. 12%) possesses many of the qualities required for a HPAC support (7) (*Table 1*). The few functional hydroxyl groups on the gel's structure may serve as potential points of attachment for various affinity ligands. Nevertheless, the applicability of cross-linked natural matrices in HPAC has yet to be demonstrated.

3. MATRIX PRE-TREATMENT: COATING, FUNCTIONALIZATION AND ACTIVATION

Silica gels *must* first be coated with appropriate organosilanes before ligand coupling. This compulsory coating process is simultaneously accompanied, depending on the organosilane used, either by matrix activation (e.g. the introduction of epoxy or isothiocyanato groups) or by matrix functionalization (e.g. the introduction of hydroxyl or amino groups). Activated silica reacts directly with ligands carrying amino, hydroxyl or sulphydryl groups, whereas functional matrices react directly only with active ligands, for example, reactive dyes. Functional hydroxyl-matrices must first be converted to an active form (e.g. aldehydyl, tresyl, imidazol carbamate or epoxy) prior to coupling

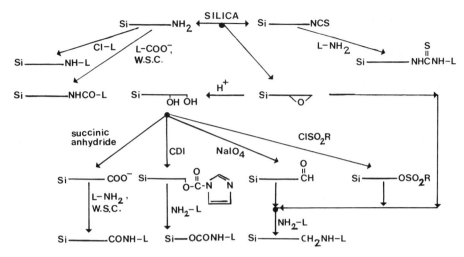

Figure 1. Chemical pathways in matrix pre-treatment and ligand coupling. W.S.C., water-soluble carbodiimide; ClSO$_2$R, tresyl chloride; CDI, 1,1′-carbonyldiimidazole; L, ligand.

to affinity ligands. However, it is possible to couple ligands and functional matrices by condensing free amino and carboxyl groups via an appropriate carbodiimide. *Figure 1* provides a schematic presentation of the chemical pathways involved in matrix pre-treatment and ligand coupling.

3.1 Glycidoxypropyl-silica (epoxy-silica): simultaneous coating and activation

3.1.1 *Preparation of epoxy-silica under aqueous conditions*

The materials required are as follows.

(i)   Spherical macroporous silica (10 g, 30 nm, 250 m^2/g).

(ii)   γ-Glycidoxypropyltrimethoxy-silane.

(iii)   0.1 M sodium acetate buffer (pH 5.5).

(iv)   1−10 mM HCl.

(v)   Acetone.

(vi)   Sintered glass funnel (porosity 4).

The method is as follows [steps (i) and (ii) apply only to newly manufactured silica].

(i)   In a 500 ml flask add 10 g of silica and 200 ml of 1−10 mM HCl. Sonicate the suspension for 10 min under reduced pressure and then heat it at 90°C for 60 min with gentle stirring using an overhead device or shaking in a bath.

(ii)   Allow the suspension to cool and wash the silica with 500 ml each of water and acetone and dry it at 140°C for 1 h.

(iii)   In a 250 ml flask add 10 g of silica and 100 ml of a 5% (v/v) solution of γ-gly-cidoxypropyltrimethoxy-silane in acetate buffer. Sonicate the suspension for 10 min under reduced pressure and heat it at 90−95°C for 5−6 h with gentle stirring as in (i).

(iv)   Allow the suspension to cool and wash the epoxy-silica with 500 ml each of water and acetone and dry it under reduced pressure. Store in a desiccator.

3.1.2 *Preparation of epoxy-silica under dry conditions (4)*

The materials required are as follows.

(i)   Spherical macroporous silica (10 g, 30 nm, 250 m^2/g).
(ii)   γ-Glycidoxypropyltrimethoxy-silane.
(iii)   Triethylamine.
(iv)   Dry toluene (sodium or molecular sieve 0.4 nm).
(v)   Toluene.
(vi)   Acetone.
(vii)   Sintered glass funnel (porosity 4).
(viii) Drying tube (CaCl$_2$ or CaSO$_4$).

The method is as follows.

(i)   In a 500 ml three-necked flask add 10 g of silica and heat the flask at 150°C for at least 6 h under vacuum (<0.01 mbar) using a vacuum pump to remove adsorbed moisture (*heat-and-vacuum* dried silica). Remember to insert a drying tube (CaCl$_2$ or CaSO$_4$) between the flask and the pump.

(ii)   Cool the flask (<80°C), disconnect the vacuum, fit with an overhead stirring device and immediately introduce a total of 150 ml of dry toluene containing already 10 ml of organosilane and 0.25 ml of triethylamine. Sonicate the suspension for 5−10 min under reduced pressure (remember to insert a drying tube between the flask and the pump).

(iii)   Disconnect the pump and immediately fit the flask with a reflux condenser carrying a drying tube at the end (both items should have been previously dried at 140°C for 90 min). Reflux the suspension for 16−18 h with gentle stirring.

(iv)   Allow the suspension to cool and then wash the epoxy-silica with 500 ml each of toluene and acetone and dry it under reduced pressure; store in a desiccator.

The amount of silane and triethylamine used should be approximately proportional to the surface area of the silica; thus, if a silica of 50 m^2/g is coated then five times less reagents should be used. The epoxy group concentration on coated silica can be determined by titration of the hydroxyl ions released during the reaction of the epoxy groups with thiosulphate (4). To do this suspend up to 100 mg of dry epoxy-silica in 2 ml water and adjust the pH to 7.0. Add 1 ml of a 3 M solution of sodium thiosulphate, pH 7.0, to initiate the reaction. During the course of the reaction (~60 min) maintain the pH at 7.0 by adding 0.1 M HCl. The consumption of HCl corresponds to the epoxy group content of the gel.

3.2 Isothiocyanatopropyl-silica: simultaneous coating and activation (8)

The materials required are as follows.

(i)   Spherical macroporous silica (5 g, typically 30 nm, 250 m^2/g).
(ii)   3-Isothiocyanatopropyltriethoxy-silane.
(iii)   Dry dichloromethane.
(iv)   Dichloromethane.
(v)   Dry 2,6-dimethylpyridine.
(vi)   Methanol.
(vii)   Diethyl ether.

(viii) Dry argon.
(ix) Sintered glass funnel (porosity 4).
(x) Drying tube.

The method is as follows.

(i) Prepare 5 g of heat-and-vacuum dried silica as described under Section 3.1.2, step (i) but extend the drying time to 12 h.
(ii) Allow the dried silica to cool to room temperature, disconnect the vacuum and immediately introduce 20 ml of dry dichloromethane containing already 12 mmol of dry 2,6-dimethylpyridine and 12 mmol of organosilane. Sonicate the suspension for 5 min under reduced pressure (remember to insert a drying tube between the flask and the pump). Disconnect the pump and immediately fit the flask with a reflux condenser carrying a drying tube at the end. Reflux under dry argon for 24 h.
(iii) Allow the suspension to cool and wash the activated silica with 250 ml each of dichloromethane, methanol and diethyl ether. Dry the silica under reduced pressure and store it in a desiccator.

3.3 Aminopropyl-silica: simultaneous coating and functionalization

Follow the procedure given in Section 3.1.1 or 3.1.2 but use 3-aminopropyltriethoxysilane as the coating agent.

3.4 Glycerylpropyl-silica (diol-silica): simultaneous coating and functionalization

3.4.1 *Preparation of diol-silica from epoxy-silica*

(i) In a 1 litre flask add 5 g of epoxy-silica (see Section 3.1) and 500 ml of $1-10$ mM H_2SO_4 or HCl. Sonicate the supsension for 5 min under reduced pressure and then heat it at $90-95°C$ for 60 min with shaking.
(ii) Allow the suspension to cool and wash diol-silica on a sintered glass funnel (porosity 4) with 250 ml each of water and acetone and dry it at $140°C$ for 60 min.

3.4.2 *Preparation of diol-silica from unmodified silica (3)*

Follow the procedure given in Section 3.1.1 but use an aqueous solution of organosilane and adjust its pH to 3.5 with HCl. The mildly acidic conditions used during the coating reaction are sufficient to convert the oxirane groups to diols.

The diol content of diol-bonded silica can be determined by oxidation with periodic acid. The iodine produced after adding KI is titrated with thiosulphate (6).

3.5 Preparation of diol-Eupergit from epoxy-Eupergit

Eupergit C 30N is a synthetic hydrophilic support carrying epoxy groups (*Table 1*). If you wish to obtain the corresponding diol form, follow the procedure below.

In a 25 ml flask add 1 g of Eupergit C 30N and 10 ml of 1 M NaOH. Leave the mixture at room temperature for 16 h. Wash the diol-Eupergit on a sintered glass funnel (porosity 4) with an excess of water until washings are neutral followed by 100 ml each of acetone/water 1:1 and acetone. Dry the gel at $40°C$ overnight.

3.6 **Aldehyde-silica: activation of diol-silica**

The materials required are as follows.

(i)   Spherical macroporous diol-silica (5 g, 30 nm, 250 m^2/g).
(ii)   Acetic acid (90%, v/v).
(iii)   Sodium periodate.
(iv)   Methanol.
(v)   Diethyl ether.
(vi)   Sintered glass funnel (porosity 4).

The method is as follows.

(i)   In a 250 ml flask add 5 g of diol-silica (see Section 3.4) and a freshly prepared solution of 100 ml 90% acetic acid containing 25 g of sodium periodate. Sonicate the suspension for 5 min under reduced pressure and shake it for 2 h at 25°C.
(ii)   Wash the aldehyde-silica with 250 ml each of water, methanol and diethyl ether. Dry the silica under reduced pressure and store it in a desiccator.

3.7 **Tresyl-silica: activation of diol-silica**

Minimum exposure to moisture is necessary since the reagent 2,2,2-trifluoroethane-sulphonyl chloride (tresyl chloride) is readily hydrolysed. The materials required are as follows.

(i)   Spherical macroporous diol-silica (5 g, e.g. 100 nm, 20 m^2/g).
(ii)   Tresyl chloride.
(iii)   Dry acetone (molecular sieve 0.4 nm).
(iv)   Acetone.
(v)   Dry pyridine (molecular sieve 0.4 nm).
(vi)   Hydrochloric acid solution (5 mM).
(vii)   Sintered glass funnel (porosity 4).

The method is as follows.

(i)   Prepare 5 g of heat-and-vacuum dried diol-silica [see Section 3.1.2, step (i) and Section 3.4]. Let the silica cool to room temperature, disconnect the vacuum and immediately proceed to the next step. Alternatively, wash the silica with 3 × 150 ml dry acetone, sucked-dry, and proceed to step (ii).
(ii)   Into the flask containing the dry diol-silica immediately add a solution made of 16 ml dry acetone and 0.5 ml of dry pyridine. Replace the stopper in the flask and cool the suspension to 0°C with gentle stirring.
(iii)   Add slowly 170 μl of tresyl chloride with vigorous stirring and when the addition has been accomplished immediately replace the stopper in the flask and stir the suspension gently for 20 min.
(iv)   Wash the tresyl-silica with 250 ml each of acetone, acetone/5 mM HCl 1:1 (v/v), 5 mM HCl and acetone. Dry the activated silica under reduced pressure and store it in a desiccator.

Tresyl-silica is stable in aqueous solution of low pH. If the activated material is to be processed the same day then the last acetone wash and drying are not necessary.

3.8 Imidazol carbamate-silica: activation of diol-silica

Dry solvents and minimum exposure to moisture are necessary since the reagent 1,1'-carbonyldiimidazole is moisture-sensitive. The materials required are as follows.

(i)   Spherical macroporous diol-silica (5 g, 50 nm, 50 m²/g).
(ii)   1,1'-Carbonyldiimidazole (CDI).
(iii)   Dry acetonitrile (molecular sieve 0.4 nm).
(iv)   Dry acetone.
(v)   Sintered glass funnel (porosity 4).
(vi)   Drying tube.

The method is as follows.

(i)   Prepare 5 g of heat-and-vacuum dried diol-silica [see Section 3.1.2, step (i) and Section 3.4]. Let the silica cool to room temperature, disconnect the vacuum and proceed with step (ii). Alternatively, wash the diol-silica with 3 × 150 ml dry acetonitrile, sucked-dry, and proceed with step (ii).

(ii)   In the flask containing the dry diol-silica (5 g) add immediately a freshly prepared solution of 40 ml of dry acetonitrile containing 2.4 g CDI. Sonicate the suspension for 6 min under reduced pressure (remember to insert a drying tube between the flask and the pump) and replace the stopper in the flask. Leave the suspension to incubate for 30 min at 25°C with gentle shaking.

(iii)   Wash the CDI-silica with 250 ml of dry acetone and dry the gel under reduced pressure. Store the activated silica in a desiccator.

3.9 Epoxy-Separon 1000: activation of a synthetic gel with free hydroxyl groups (9)

By varying the amount of epichlorohydrin added (e.g. 40, 25 or 1.6 ml/10 g gel) matrices activated to different extents are obtained, approximately 1770, 800 and 140 μmol/g, respectively. The materials required are as follows.

(i)   Separon H1000 (see *Table 1*).
(ii)   Epichlorohydrin.
(iii)   Potassium hydroxide (50%).
(iv)   Aqueous potassium bromide (saturated).
(v)   Acetone.
(vi)   Diethyl ether.
(vii)   Sintered glass funnel (porosity 4).

The method is as follows.

(i)   Leave 10 g of Separon in 50 ml of 50% (w/v) KOH solution for 15 h at room temperature.

(ii)   Filter the gel on a glass filter by suction and transfer it to a 250 ml flask containing 40 ml epichlorohydrin (for maximum activation). After about 60 min the temperature of the suspension rises to the boiling point of the activating agent. Add 10 ml of saturated aqueous KBr solution and boil the suspension with stirring for 3 h.

(iii)   Allow the suspension to cool and wash the epoxy-Separon on a glass filter three times with 100 ml of acetone/water until the washings are neutral, three times with 100 ml acetone/diethyl ether. Dry the activated matrix under reduced pressure and store in a desiccator.

3.10 Gels with free carboxyl groups: another functional form of HPLC matrices

Here two procedures are described by which one can introduce free carboxyl groups on with hydroxyl groups. In both cases one must use anhydrous conditions and dry solvents (molecular sieve 40 nm).

3.10.1 *Succinyl-silica (10)*

The materials required are as follows.

(i)   Spherical macroporous diol-silica (5 g, 50 nm, 50 m^2/g).
(ii) Succinic anhydride.
(iii) Dry dioxane.
(iv) Dioxane.
(v)   Sintered glass funnel (porosity 4).
(vi) Drying tube.

The method is as follows.

(i)   Prepare 5 g of heat-and-vacuum dried diol-silica [see Section 3.1.2, step (i) and Section 3.4]. Let the silica cool to room temperature, disconnect the vacuum pump and immediately proceed with step (ii). Alternatively, wash the diol-silica with 3 × 150 ml dry dioxane, sucked-dry, and proceed with step (ii).
(ii) In the flask containing the dry diol-silica (5 g) add a freshly prepared solution of 250 ml dry dioxane containing 1.3 g of succinic anhydride. Sonicate the suspension for 5 min under reduced pressure (remember to insert a drying tube between the flask and the pump). Replace the stopper in the flask and shake the suspension for 24 h at 25°C.
(iii) Wash the succinyl-silica with alternating warm and room temperature portions of dioxane (6 × 50 ml) and dry the gel under reduced pressure.

3.10.2 *TSK-acetylglycylglycine (TSK-AGG): a synthetic gel with an ω-carboxyl spacer (11)*

The materials required are as follows.

(i)   TSK gel with free hydroxl groups (e.g. Toyopearl HW65S or TSK PW-type).
(ii)   Methylsulphinyl carbanion (ref. 12).
(iii) *N*-Chloroacetylglycylglycine (CAGG) (ref. 13).
(iv)   Sodium borohydride.
(v)   Dry acetone.
(vi)   Acetone.
(vii) Water/acetone, 1:1 (v/v).
(viii) Dry dimethylsulphoxide (DMSO).
(ix)   DMSO.
(x)   0.05 M HCl.
(xi)   0.05 M NaOH − 1 M NaCl.

(xii)   Dry nitrogen.

(xiii) Sintered glass funnel (porosity 4).

The method is as follows.

(i)   Wash a portion of the TSK gel on a glass filter with water, water/acetone (1:1), acetone and dry acetone. Dry the gel under reduced pressure using a drying tube between the flask and the pump. Store in a desiccator.

(ii)   In a flask containing 10 g of dry TSK gel add a freshly prepared solution of 80 ml of dry DMSO containing 1.0 g of sodium borohydride. Immediately supply the flask with a stream of dry nitrogen and sonicate for 5 min.

(iii)   Add 9 ml of a 2.25 M solution of methylsulphinyl carbanion (12) in dry DMSO and leave the suspension for 5 min with gentle shaking.

(iv)   Add 8 ml of dry DMSO containing 4 mmol of CAGG (13) and leave the reaction for 60 min at 25°C with occasional shaking.

(v)   Transfer the reaction mixture into 1 litre of ice-cold water and then wash the TSK-AGG gel on a glass filter with 1 litre each of water, 0.05 M HCl, 0.05 M NaOH − 1 M NaCl, water until the washings are neutral, water/acetone (1:1) and acetone. Dry the gel at 40°C overnight.

All methods described above for activating gels with hydroxyl groups as well as for introducing free carboxyl groups to hydroxyl-gels can be applied, in general, to any type of HPLC matrix with functional hydroxyl groups.

4. LIGAND COUPLING TO FUNCTIONAL AND ACTIVATED MATRICES

4.1 Ligand coupling to functional matrices

4.1.1 *Coupling of dichlorotriazinyl-dyes to diol-silica (14)*

The materials required are as follows.

(i)   Spherical macroporous diol-silica (1 g, 30 nm, 250 m^2/g).

(ii)   Procion blue MX-R or any other dichlorotriazinyl-dye.

(iii)   0.1 M NaHCO$_3$.

(iv)   Dimethylsulphoxide (DMSO).

(v)   Acetone.

(vi)   Sintered glass funnel (porosity 4).

The method is as follows.

(i)   In a 50 ml flask add 200 mg of dye, 1.5 ml DMSO, 8.5 ml NaHCO$_3$ solution and immediately sonicate the mixture for 30 sec. Add 1 g of diol-silica (see Section 3.4), sonicate for 5 min under reduced pressure and allow the suspension to incubate at 37°C for 20 h with shaking.

(ii)   Wash the blue-silica with water until the washings are dye-free followed by 100 ml acetone. Dry the gel at 90°C for 60 min.

The concentration of the immobilized ligand should be approximately 8.5 μmol/g dry gel; molar absorption coefficient for Procion blue MX-R in glycerol/H$_2$O (1:1, v/v) equals 11 600 l/mol/cm. For silica gels the pH should never exceed 8.5 at a maximum temperature of 40°C and even then for a few hours only. For this reason, the performance of the above method is poor when coupling monochlorotriazinyl-dyes which require higher pH and temperature. For efficient coupling of monochlorotriazinyl-dyes the procedures given in Sections 4.1.2 and 4.2.1 are recommended.

4.1.2 *Coupling of chlorotriazinyl-dyes to amino-silica*

Follow the procedure given in Section 4.1.1 using amino-silica (see Section 3.3) and any chlorotriazinyl-dye. The concentration of immobilized ligand could be up to 25 μmol/g dry gel.

4.1.3 *Coupling of chlorotriazinyl-dyes to synthetic gels with functional hydroxyl groups (14)*

The materials required are as follows.

(i)   Dry-form gel; TSK PW-type (see Section 2.2.2) or diol-Eupergit C 30N (see Sections 2.2.1 and 3.5).

(ii)   Procion blue MX-R or any other dichlorotriazinyl-dye.

(iii)   1% (w/v) Na_2CO_3.

(iv)   Dimethylsulphoxide (DMSO).

(v)   Acetone.

(vi)   Acetone/H_2O, 1:1 (v/v).

(vii)   Sintered glass funnel (porosity 4).

The method is as follows.

(i)   In a 50 ml flask add 100 mg of dye, 4 ml of DMSO, 21 ml of Na_2CO_3 solution and immediately sonicate the mixture for 30 sec. Add 1 g of gel, sonicate for 5 min under reduced pressure and allow the suspension to incubate at 40°C for 14 h with shaking.

(ii)   Wash the blue-gel with water until the washings are dye-free, 100 ml acetone/H_2O (1:1), 200 ml acetone and dry it at 40°C overnight.

This procedure should yield adsorbents with $12-15$ μmol dye/g dry gel. However, the use of monochlorotriazinyl-dyes under the conditions described above inevitably leads to lower ligand substitution on the gel. Thus, for this type of less reactive dye a higher temperature and/or longer incubation time is needed. Alternatively, the monochlorotriazinyl-dye can be converted to the ω-aminoalkyl-analogue (Section 4.2.1) and then coupled to previously activated gels with hydroxyl groups (Sections 3.5, 3.7 or 3.8). The latter procedure affords gels with high concentration of immobilized dye.

4.1.4 *Coupling of ligands carrying primary amino groups to gels with free carboxyl groups: application to the coupling of trimethyl(p-aminophenyl)ammonium chloride and p-aminobenzamidine to ACA-Separon (16).*

The materials required are as follows.

(i)   ACA-Separon (Laboratory Instruments Works, Prague, Czechoslovakia).

(ii)   Trimethyl(*p*-aminophenyl)ammonium chloride or *p*-aminobenzamidine (ligands).

(iii)   1-Ethyl-3-(3-dimethylaminopropyl)carbodiimide hydrochloride.

(iv)   0.05 M Sodium hydroxide solution-1 M NaCl.

(v)   Acetone.

(vi)   Water/acetone, 1:1 (v/v).

(vii)   Sintered glass funnel (porosity 4).

The method is given below.

(i) In a 100 ml flask add in the following order: a freshly prepared solution of 500 mg of carbodiimide in 35 ml water (pH 5.0), 270 mg of ligand and 5 g of ACA-Separon (Separon substituted with a spacer carrying a terminal free carboxyl group). Sonicate the suspension under reduced pressure for 5 min and leave to incubate at 25°C for 24 h with shaking.

(ii) Wash the gel with 500 ml each of NaOH − NaCl solution, water, water/acetone (1:1), acetone and leave to dry at 40°C overnight. Immobilized-ligand concentration of approximately 40 μmol/g is expected.

The above carbodiimide-promoted condensation can also take place in an aqueous buffer of 0.2 M 2-(morpholino)ethanesulphonic acid−NaOH (Mes−NaOH), pH 4.7, with periodic adjustment of the pH with 1 M HCl or 1 M NaOH (12).

4.2 Ligand coupling to active matrices

Ligands carrying amino, sulphydryl or hydroxyl groups can be coupled to matrices activated as described earlier (Sections 3.1, 3.2, 3.6−3.9). Here only some typical examples are given but, obviously, the methods can be applied to any appropriate ligand and matrix.

4.2.1 *Coupling of ω-aminoalkyl-triazinyl-dyes to activated silicas*

(i) *Preparation of an ω-aminoalkyl-triazinyl-dye (6-aminohexyl-Cibacron blue F3G-A) (16).* The following materials are required.

(a) Chlorotriazinyl-dye (Cibacron blue F3G-A).
(b) α,ω-Diaminoalkane (a solution of 1,6-diaminohexane, 1 M, pH 10; 5 M HCl).
(c) 0.3 M HCl.
(d) Acetone.
(e) Diethyl ether.
(f) Paper filter (Whatman, hardened).

The procedure is as follows.

(a) In a 100 ml flask add 20 ml of water, 10 ml of a solution of 1,6-diaminohexane (1 M, pH 10; 5 M HCl), 1.3 g (~1 mmol) of Cibacron blue F3G-A (~60% pure) and heat the mixture at 50°C for 90 min under stirring.

(b) Introduce the reaction mixture dropwise (you could use a peristaltic pump at a flow-rate of 60 ml/h) in 200 ml of 0.3 M HCl with gentle stirring. Allow the precipitate formed to settle for 90 min at room temperature.

(c) Wash the precipitated (6-aminohexyl-Cibacron blue F3G-A) on a paper filter with 200 ml each of 0.3 M HCl, acetone until washings are dye-free, diethyl ether and dry the product at 40°C overnight.

(ii) *Coupling of an ω-aminoalkyl-triazinyl-dye (6-aminohexyl-Cibacron blue F3G-A) to epoxy-silica or isothiocyanato-silica (8,15).* The following materials are required.

(a) Activated silica (5 g, see Sections 3.1 and 3.2).
(b) 6-Aminohexyl-triazinyl-dye [see Section 4.2.1(i)].
(c) Sodium bicarbonate, 0.1 M.

(d)   Acetone.

(e)   Sintered glass funnel (porosity 4).

The method is as follows.

(a)   In a 100 ml flask add 40 ml of 0.1 M sodium bicarbonate solution, 500 mg of 6-aminohexyl-Cibacron blue F3G-A and sonicate for 1 min. Add 5 g of epoxy-silica or 3-isothiocyanato-silica, sonicate the suspension for 5 min under reduced pressure and incubate at 35°C for 12 h with shaking.

(b)   Wash the dyed silica gel with water until the washings are dye-free followed by 500 ml acetone and dry the gel at 90°C for 60 min. Immobilized dye concentration of $6-8$ μmol/g dry gel is expected.

4.2.2 Coenzyme and nucleotide analogues: coupling of N^6-substituted NAD^+ and AMP to activated silicas

(i) *Coupling of N^6-substituted NAD to epoxy-silica (4).* The following materials are required.

(a)   Epoxy-silica (1 g, see Section 3.1).

(b) N^6-[N-(6-aminohexyl) carbamylmethyl]-NAD (available from Sigma).

(c)   0.1 M Sodium pyrophosphate buffer (pH 8.0).

(d) H_2SO_4 (pH 2.0).

(e)   Ethanol.

(f)   Diethyl ether.

(g)   Sintered glass funnel (porosity 4).

The method is as follows.

(a)   In a 10 ml flask add in the following order: 3 ml of buffer, 20 mg of N^6-substituted NAD and 1 g of epoxy-silica. Sonicate the suspension for 5 min under reduced pressure and incubate at 20°C for 5 days.

(b)   Wash the NAD^+-silica with H_2O until the washings are neutral. Suspend the silica in 100 ml of H_2SO_4 (pH 2.0) and heat at 50°C for 4 h.

(c)   Wash the NAD^+-silica with 100 ml each of water, ethanol, diethyl ether and dry the gel under reduced pressure. Immobilized-NAD^+ concentration of approximately 1 μmol/g dry silica is expected.

(ii) *Coupling of N^6-substituted AMP to aldehyde-silica (17).* The following materials are required.

(a)   Aldehyde-silica (1 g, see Section 3.6).

(b) N^6-(6-aminohexyl)-AMP (see ref. 18).

(c)   0.1 M Sodium bicarbonate.

(d)   Sodium borohydride.

(e)   Ethanol.

(f)   Diethyl ether.

(g)   Sintered glass funnel (porosity 4).

The method is as follows.

(a)   In a 10 ml flask add in the following order: 2.5 ml of sodium bicarbonate solution, 200 mg of N^6-substituted AMP and 1 g of aldehyde-silica. Sonicate the suspen-

sion for 5 min under reduced pressure and incubate at 25°C for 24 h with shaking.
(b)   Prepare a solution of 40 mg of sodium borohydride in 0.5 ml water and add it, at intervals of 60 min, in the silica reaction mixture with shaking.
(c)   Wash the AMP-silica with 100 ml each of water, ethanol, diethyl ether and dry the gel under reduced pressure. Immobilized-AMP concentration of approximately 20 μmol/g dry silica is expected.

4.2.3 *Protein and enzyme coupling to activated gels*

(i) *Coupling of concanavalin A and mannan to isothiocyanato-silica (8)*. The following materials are required.
(a)   Isothiocyanoto-silica (5 g, see Section 3.2).
(b)   Concanavalin A (Con A) or mannan.
(c)   0.15 M Sodium phosphate buffer (pH 8.0).
(d)   Sintered glass funnel (porosity 4).

The method is as follows.

(a)   In a 250 ml flask add in the following order: 70 ml of phosphate buffer, 170 mg of Con A or 420 mg mannan and 5 g of isothiocyanato-silica. De-gas the suspension under vacuum with occasional shaking and incubate at 20°C for 24 h with gentle shaking.
(b)   Wash the protein-silica with 500 ml of water and 500 ml of the solvent system to be used afterwards. Store the gel wet at 4°C. The amount of protein immobilized should be approximately 30 mg/g of dry silica.

(ii) *Coupling of anti-human serum albumin to aldehyde-silica (17)*. The following materials are required.
(a)   Aldehyde-silica (5 g, see Section 3.6).
(b)   Anti-human serum albumin (anti-HSA).
(c)   Sodium borohydride.
(d)   0.1 M Sodium bicarbonate buffer (pH 7.9) (buffer A).
(e)   Buffer A containing 0.5 M NaCl (buffer B).
(f)   Sintered glass funnel (porosity 4).

The method is as follows.

(a)   In a 50 ml flask add in the following order: 10 ml of buffer A, 80 mg of anti-HSA in 4 ml of buffer A and 5 g of aldehyde-silica. De-gas the suspension under reduced pressure and leave at 4°C for 50 h.
(b)   In the suspension add portions from a solution of 160 mg of sodium borohydride in 1.6 ml of water at a total period of 16 h.
(c)   Wash the anti-HSA-silica with 500 ml of buffer B and 500 ml of the solvent system to be used afterwards. Store the gel wet at 4°C.

(iii) *Coupling of alcohol dehydrogenase to tresyl-silica (4)*. The materials required are as follows.
(a)   Tresyl-silica (1 g, see Section 3.7).
(b)   Alcohol dehydrogenase (ADH).

(c)   0.4 M Sodium phosphate buffer (pH 7.0) containing 0.2 M isobutyramide (buffer A).

(d)   Adenine nicotinamide; NADH.

(e)   0.2 M Tris−HCl buffer (pH 8.0) containing 1 mM dithioerythreitol (buffer B).

(f)   0.1 M Sodium phosphate buffer (pH 7.5) containing 0.5 M NaCl and 1 mM dithioerythreitol (buffer C).

(g)   0.1 M Sodium phosphate buffer (pH 7.5) containing 1 mM dithioerythreitol (buffer D).

(h)   Sintered glass funnel (porosity 4).

The method is outlined below.

(a)   In a 25 ml flask add 3 ml of buffer A containing 2 mM NADH and 1 g of tresyl-silica. De-gas the suspension under reduced pressure for 5 min and then add 3 ml of buffer A containing 20 mg ADH; leave at room temperature for 20 h.

(b)   Wash the ADH-silica with 100 ml of buffer B and transfer the gel in 10 ml of buffer B. Leave to incubate for 60 min at room temperature.

(c)   Wash the ADH-silica with 200 ml of buffer C, 100 ml of buffer D and store it at 4°C. The amount of ADH immobilized should be approximately 20 mg/g of dry silica.

(iv) *Coupling of concanavalin A to imidazol carbamate-Dynospheres*. The materials required are as follows.

(a)   Imidazol carbamate-Dynospheres XP-3507 (activation is performed as for silica in Section 3.8).

(b)   Concanavalin A.

(c)   50 mM Phosphate buffer, pH 7.0 (buffer A).

(d)   Buffer A containing 1 M NaCl (buffer B).

(e)   Sintered glass funnel (porosity 4).

The method is as follows.

(a)   In a 50 ml flask add 2 g of activated gel (moist) followed by 8 ml of buffer A containing 100 mg of Con A (Sigma C-7275). Leave the suspension to react at room temperature for 24 h with gentle shaking.

(b)   Wash the Con A-Dynospheres particles with 200 ml each of cold buffers A and B. Store particles moist at 4°C. During prolonged storage it is recommended to use 0.02% (w/v) sodium azide to prevent biological contamination.

4.2.4 *Sugar analogues: coupling of p-aminophenyl-α-D-mannopyranoside to imidazol carbamate-silica (10)*

The materials required are listed below.

(i)   Imidazol carbamate-silica (5 g, see Section 3.8).

(ii)   *p*-Aminophenyl-α-D-mannopyranoside (*p*-APM).

(iii)   0.1 M Sodium phosphate buffer (pH 7.0).

(iv)   2 M NaCl.

(v)   Ethanol.

(vi)   Diethyl ether.

(vii)   Sintered glass funnel (porosity 4).

(viii) Nitrogen.

The method is outlined below.

(i)   In a 50 ml flask add in the following order: 20 ml of phosphate buffer, 50−400 mg of *p*-APM and 5 g of activated-silica. Sonicate the suspension under reduced pressure for 10 min, flush with nitrogen and incubate at 25°C for 48 h under shaking.

(ii)   Wash the *p*-APM-silica with 500 ml of 2 M NaCl, 200 ml of water, ethanol and diethyl ether. Dry the gel under reduced pressure. Immobilized ligand concentrations of $0.3 - 1.0$ μmol/m^2 are expected.

4.2.5 *Protease inhibitors: coupling of ε-aminocaproyl-L-phenylalanyl-D-phenylalanine methyl ester to epoxy-Separon 1000 (9)*

The following materials are required.

(i)   Epoxy-Separon 1000 (5 g, see Section 3.9).
(ii)   Amino-tripeptide hydrochloride.
(iii)   Dimethylformamide (DMF).
(iv)   Triethylamine.
(v)   Ethanol.
(vi)   Diethyl ether.
(vii)   Sintered glass funnel (porosity 4).

The method is outlined below.

(i)   In a 100 ml flask add in the following order: 25 ml of DMF, 125 μl of triethylamine, 0.3 g of amino-tripeptide hydrochloride and 5 g of epoxy-Separon (maximum activation). Sonicate the suspension for 5 min under reduced pressure and incubate at 25°C for 48 h with shaking.

(ii)   Wash the gel with 500 ml each of DMF, water, ethanol and diethyl ether. Dry the gel under reduced pressure. The amount of ligand immobilized should be approximately 2 μmol/g wet gel.

5. DETERMINATION OF IMMOBILIZED LIGAND CONCENTRATION

Characterization of an affinity matrix requires the determination of immobilized ligand concentration. This measurement is made in different ways depending on the nature both of the ligand bound and the gel. When the ligand absorbs at a particular wavelength, then quantitation is readily achieved spectrophotometrically using the molar absorption coefficient of the free ligand. In order to measure the absorbance of the ligand immobilized, one can first solubilize the gel under appropriate conditions and then measure the absorbance of the solution. This method can be applied to silica-based gels which are readily soluble in alkaline conditions. However, for synthetic or cross-linked natural gels whose three-dimensional structure cannot easily be destructed, the usual approach is to suspend the gel intact in a viscous liquid, for example glycerol/H$_2$O (1:1, v/v), and measure the absorbance of the suspension at the particular λ_{max} of the ligand. Alternatively, the bound ligand can be determined by the difference of the amounts added in the coupling reaction and that recovered in the gel washings after the end of the reaction; however, this method is not reliable, nevertheless, it is used for protein and enzyme ligands. For ligands, which exhibit no absorption characteristics, quantitation is achievable by elementary analysis or specific tests for particular chemical groups present on the ligand.

5.1 **Determination by gel solubilization (silica matrices only) (15)**

Weigh a dried silica gel sample (15−60 mg), add it to 5 ml of 1 M NaOH and heat the suspension at 60°C for 30 min to solubilize the silica. Dilute the solution to about 25 ml with water, adjust the pH to 7.0 with 1 M HCl, add 5 ml of 1 M potassium phosphate buffer (pH 7.5) and make the solution up to 50 ml with water. Read the absorbance of the solution at the particular λ_{max} of the ligand against plain silica of the same weight and treated in the same way.

5.2 **Determination by gel suspension (14)**

In a vial add the sample (∼3 mg) of dried gel and 3 ml of 50% (v/v) glycerol in water. Sonicate the suspension for 30 sec and allow to stand for 10 min. Shake the suspension gently a few times and immediately read the absorption at the particular λ_{max} of the ligand against a suspension of plain gel of the same weight treated under the same conditions.

For silica-based matrices it is better to use 100% dimethylsulphoxide (DMSO) or 50% (v/v) glycerol in DMSO. Silica has the same refractive index as DMSO so less light is scattered when using DMSO instead of water. Remember to use figures of molar absorption coefficients which are determined in liquid of the same composition as that used for gel measurements.

6. PACKING OF HPAC GELS

A very good way of packing HPAC columns is the slurry method. For that method the following accessories are needed.

(i) Compressed-nitrogen cylinder.
(ii) Air-driven liquid pump [e.g. Haskel, model MCP-71, maximum pressure output 8×10^3 p.s.i. (500 bar), minimum pressure input 25 p.s.i. (1.5 bar)].
(iii) Manometer (maximum input 10×10^3 p.s.i., 700 bar).
(iv) High-pressure two-way HPLC valve.
(v) High-pressure HPLC filter (maximum pressure 15×10^3 p.s.i., 10^3 bar, 0.5 μm pore size).
(vi) Packing bomb (e.g. 70 ml capacity, ICAM, Clwyd, UK).
(vii) Empty stainless steel columns (e.g. 6 mm o.d.) complete with matching end fittings, frits (maximum pore size 2 μm) and of appropriate outer diameter to fit to the packing bomb. If necessary an adaptor can be used to match column and packing bomb.

A complete set-up similar to the above can be obtained as 'Self Assembly Kit for Packing HPLC Columns' (e.g. Phase Sep, Clywd, UK). When packing a gel into a column usually it is best to start at a low pre-set pressure for a few minutes and then increase the pressure to 70−80% of the maximum permitted pressure for the particular gel. The time required to completely pack a given bed volume relates inversely to the slurry concentration and flow-rate achieved at the pressure applied. The amount of gel required

Table 2. Packing conditions of HPAC gels.

| Gel type (beaded form) | Particle size (μm) | Starting conditions | | Final conditions | | Back-pressure developed at 1 ml/min in water (bar) |
		Pressure (bar)	Time (min)	Pressure (bar)	Time (min)	
Silica	10−12	80	10	140	90	60
TSK-PW	10	60	10	100	30	25
Eupergit C 30N	30	45	3	80	7	0

Pressure figures refer to maximum pressure readings as shown by the manometer (1 bar = 14.5 p.s.i. = 0.1 MPa). The column bed was 0.45 × 5.0 cm and the solvent was filtered water (0.45 μm). The slurry concentration was 1% (w/v), but concentrations of up to 10% (w/v) may be used to reduce packing time.

to pack a column can be estimated approximately using a packing density of 0.30−0.35 g dry gel/ml for the gels shown in *Table 2*. Here a general method for packing HPL(A)C gels in water under conditions specific for each type of gel is described (see *Table 2*) (15).

(i)   Equilibrate the pump in water by passing about 100 ml of liquid. Close the HPLC valve and set pressure to the 'starting' value using the nitrogen cylinder valve.

(ii) Suspend the required amount of dry gel in 60 ml of filtered water (0.45 μm) and add the suspension along with a magnetic bar in the packing bomb. Close the lid and tighten up all screws very well using a spanner.

(iii) Place the packing bomb on a magnetic stirrer and, while the HPLC valve remains closed (no pressure delivered at the outlet), connect the outlet from the pump to the packing bomb; use a spanner. Start stirring the suspension at about 250 r.p.m.

(iv) Connect an empty stainless steel column at the packing bomb and tighten the column with a spanner. Remember to remove the frit only from the side of the column which is connected to the bomb. Fit some tubing to the column outlet so that it is possible to collect in a measuring cylinder the water effluents during the packing.

(v)   Slowly release starting pressure by opening the HPLC valve. After a few minutes have passed and while the column is still packing under 'starting pressure', increase packing pressure to the 'final pressure' using the nitrogen cylinder valve. Leave column to pack completely for the time required (*Table 2*).

(vi) When packing is completed, close the HPLC valve (no more pressure is delivered to packing bomb or column), stop stirring the suspension, leave column for 2−4 min to almost reach the atmospheric pressure (no liquid flow from the outlet) and carefully disconnect the column from the packing bomb.

(vii) Level off packing material with the edge of a spatula and place the frit and end fitting. Label this side as 'column inlet'.

7. APPLICATIONS OF HPAC

Although high-performance affinity chromatography is a relatively new techique it is already being used in an increasing number of applications for the resolution and purifi-

Table 3. Applications of HPAC in the resolution and purification of biomolecules.

Immobilized ligand	Interacting species
Antibodies	
Anti-IgG	IgG
Anti-human serum albumin	Human albumin
Anti-bovine insulin	Insulin
Anti-creatine kinase	Creatine kinase
Dyes and dye analogues	
Cibacron blue F3G-A	Dehydrogenases, kinases, ribonuclease A
Procion blue MX-R	Lactate dehydrogenase
Procion green MX-5BR	Hexokinase
Procion brown MX-5BR	Tryptophanyl-tRNA synthetase
Procion yellow H-A	Carboxypeptidase
Procion orange H-ER	Human fibroblast extract
Procion red H-8BN	Alkaline phosphatase
Nucleotides and coenzymes	
AMP	Alcohol and lactate dehydrogenases
NAD$^+$	Lactate dehydrogenase
Proteins	
Alcohol dehydrogenase	Nucleotides
Bovine serum albumin	D,L-amino acids
Concanavalin A	Carbohydrates, glycoenzymes, horseradish peroxidase
Protein A	Immunoglobulins
Miscellaneous	
p-Aminobenzamidine	Plasmin, plasminogen, acetyl-cholinesterase, trypsin
ε-Aminocaproyl-L-Phe-D-Phe-OCH$_3$	Pepsin, proteinase
Boronic acid	Nucleotides, carbohydrates
Glucosamine	Concanavalin A
Lysine	Peroxidase, plasminogen
Mannan	Human plasma membrane proteins
Phenothiazine-analogue/Melittin	Calmodulin
Soyabean trypsin inhibitor	Trypsin, chymotrypsin
Thymine	Nucleic acid analogues
Zinc/iminodiacetic acid	Ribonuclease A, transferrin, carbonic anhydrase, lactoferrin, venom phospholipase

cation of biological molecules, especially enzymes and proteins. *Table 3* summarizes various applications of HPAC in the separation technology of biomolecules. All buffers and protein samples employed in HPAC runs should be filtered through 0.2 μm hydrophilic membrane prior to use (e.g. Millipore GV or WP type).

7.1 Purification of dehydrogenases: application to rabbit muscle L-lactate dehydrogenase

Procion blue MX-R is a competitive inhibitor with NAD$^+$ for lactate dehydrogenase. The immobilized form of this dye is used to purify lactate dehydrogenase from a crude

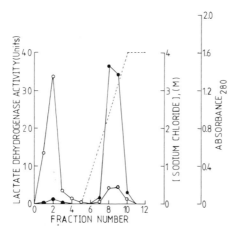

Figure 2. Purification of rabbit muscle L-lactate dehydrogenase (LDH) on TSK G5000PW-Procion Blue MX-R. HPAC column, TSK G5000PW-Procion blue MX-R (13 μmol/g dry gel), 5.0 cm × 0.45 cm; sample applied, 1 ml, 100 units LDH, 12.3 mg protein; flow-rate, 1 ml/min; starting solvent, 0.05 M sodium phosphate buffer (pH 7.0); eluant, linear gradient of NaCl (10 ml; 0−4 M) in buffer. Collected 2 ml fractions. (● − ●) enzyme activity, (○ − ○) protein; A_{280}, (----) NaCl gradient.

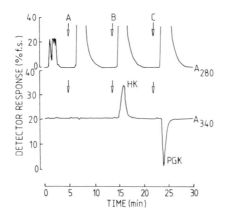

Figure 3. Separation of yeast hexokinase (HK) and 3-phosphoglycerate kinase (PGK), on silica-bound 6-aminohexyl-Cibacron blue F3G-A. HPAC column, silica-bound 6-aminohexyl-Cibacron blue F3G-A (~6 μmol/g dry gel), 10 cm × 0.5 cm; sample applied, 2 μl crude yeast extract; flow-rate, 1 ml/min; starting solvent, 0.1 M Tris−HCl buffer (pH 7.3) containing 0.5 mM EDTA, 5 mM $MgCl_2$ and 0.5 mM 2-mercaptoethanol; eluants (in buffer), **A**, 10 mM Mg.ATP; **B**, 10 mM Mg.ATP plus 25 mM D-glucose (or D-mannose); for eluting HK activity, and **C**, 10 mM Mg.ATP plus 3-phosphoglycerate; for eluting PGK activity. HK activity was detected by following the absorbance increase, whereas PGK activity was detected by following the absorbance decrease (340 nm). The total protein was monitored by the absorbance at 280 nm. Reproduced from (16) with permission.

rabbit muscle extract. When a 1 ml sample (100 units, 12.3 mg protein) of dialysed rabbit muscle crude extract is loaded on a Blue MX-R-TSK G5000 PW column (13 μmol/g dry gel) (see Section 4.1.3) most proteins passed through unbound whereas lactate dehydrogenase is adsorbed by the blue gel and is subsequently eluted by a linear gradient of NaCl (*Figure 2*). This HPAC method affords lactate dehydrogenase purified

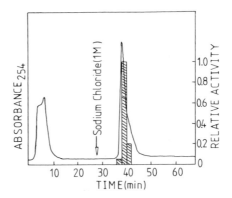

Figure 4. Purification of eel acetylcholinesterase on ACA-Separon-immobilized trimethyl (*p*-aminophenyl) ammonium chloride. HPAC column, ACA-Separon-immobilized TAPA (40 μmol/g dry gel), 25 cm × 0.8 cm; sample applied, 0.5 ml eel extract containing about 35 units enzyme and 9.5 mg protein; flow-rate, 1 ml/min; starting solvent, 0.01 M sodium phosphate buffer, (pH 7.2); eluant, 1 M NaCl in buffer, (—) protein; A_{254}, hatched regions indicate enzyme activity. Reproduced from (18) with permission.

5.6 times with about an 80% yield. The protein content is measured by the $A_{280/260}$ method; a 0.1% solution of pure LDH gives an absorbance of 1.44 at 280 nm. The protein sample and all solutions should be filtered through a 0.2 μm hydrophilic filter (Durapore) before use. TSK G5000PW gel can be obtained from Toyo Soda, Japan.

7.2 Separation of kinases: application to yeast hexokinase and 3-phosphoglycerate kinase

Cibacron blue F3G-A displays affinity for a wide range of nucleotide-binding enzymes. This has been exploited in the resolution of two kinases from a crude yeast extract. The HPAC column, 6-aminohexyl-Cibacron blue F3G-A-silica (see Section 4.2.1), is loaded with a crude yeast extract which contains the two enzymes hexokinase and 3-phosphoglycerate kinase and which are both adsorbed on the column. Unretarded proteins are washed off and subsequently the two enzymes are eluted separately, the first in the presence of 10 mM Mg.ATP and 25 mM D-glucose or D-mannose and the second in the presence of 10 mM Mg.ATP and 3-phosphoglycerate (*Figure 3*).

7.3 Purification of serine esterases and proteases: application to eel acetyl-cholinesterase and human serum plasminogen

Esterases and proteases represent a large group of hydrolytic proteins including serine and sulfhydryl enzymes. The affinity of serine esterases in binding certain cationic ligands has been exploited in the purification of these enzymes by HPAC. For example, trypsin and urokinase bind *p*-aminobenzamidine whereas acetylcholinesterase binds trimethyl(*p*-aminophenyl)ammonium chloride (TAPA). Accordingly, when eel acetylcholinesterase (0.5 ml, 35 units, 9.5 mg protein) is applied on a column containing ACA-Separon-immobilized TAPA (see Section 4.1.4) (16) inert proteins appear in the washings whereas acetylcholinesterase is adsorbed by the affinity gel. Elution of acetylcholinesterase is subsequently effected with 1 M NaCl in the irrigating buffer

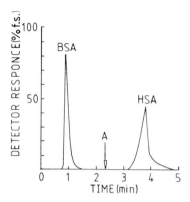

Figure 5. Separation of human serum albumin (HSA) from bovine serum albumin (BSA) on anti-HSA-silica. HPAC column, anti-HSA-silica (4 mg/g), 10 cm × 0.5 cm; sample applied, a mixture of HSA and BSA (each ~50 μg); flow-rate, 1.5 ml/min; starting solvent, 0.1 M sodium phosphate buffer (pH 7.5); eluant (A), glycine−HCl buffer, 0.2 M, pH 2.2. The protein content was monitored at 280 nm. Reproduced from (17) by permission.

(*Figure 4*). This affinity technique affords enzyme purified 38 times at about 100% yield (16). ACA-Separon can be obtained from Laboratory Instruments Works, Prague, Czechoslovakia.

In a similar fashion, human serum plasminogen binds to *p*-aminobenzamidine immobilized onto Toyopearl® HW65S via an acetylglycylglycine-spacer (see Section 3.10.2). Plasminogen (10 μg) is adsorbed by the affinity gel in 0.05 M sodium phosphate buffer, (pH 7.4) containing 0.1 M NaCl, and is subsequently eluted by 0.02 M 6-aminohexanoic acid in the above buffer system (11).

7.4 Purification by immuno HPAC: application to human serum albumin (HSA)

The technique of HPAC has been exploited in immunoadsorbent separations. Anti-HSA is immobilized to aldehyde-silica and the resultant anti-HSA-silica gel is used to separate HSA from BSA. When a sample containing HSA and BSA is applied on the HPAC immunoadsorbent, BSA passed through unbound (>98%), whereas HSA adsorbed and subsequently eluted (95% recovery) by lowering the pH to 2.2 (*Figure 5*) (17).

7.5 Resolution by metal chelate-type HPAC: application to ribonuclease A, transferrin and carbonic anhydrase

This technique, introduced by Porath *et al.* in 1975, can now be applied to enzyme resolution and purification in its HPAC version. The gel employed is microparticulated (10 μm) bearing the ligand iminodiacetic acid which is capable of chelating with metal-ions, for example, zinc. The gel, under the trade-name TSK gel Chelate-5PW (Toyo Soda, Japan) can resolve a mixture of ribonuclease A (bovine pancreas), transferrin (human) and carbonic anhydrase (bovine erythrocytes) in 0.02 M Tris−HCl buffer (pH 8.0) containing 0.5 M NaCl, after the application of 0−0.2 M glycine gradient. Prior to sample loading the column is treated with 0.2 M $ZnCl_2$ (19).

7.6 Resolution and purification by hydrophobic-type HPLC at a preparative scale

Hydrophobic-type liquid chromatography has been widely used in enzyme and protein purification technology and now provides rapid resolution and purification in its HPLC form. A microparticulate beaded support consisting of an entirely synthetic hydrophilic porous matrix bearing a hydrophobic phenyl-ligand is now available as TSK gel Phenyl-5PW (Toyo Soda, Japan). This adsorbent is capable, for example, of resolving a protein mixture in the following order: myoglobin, ribonuclease A, lysozyme, α-chymotrypsinogen and α-chymotrypsin in 0.1 M phosphate buffer (pH 7.0) after the application of a $1.8-0$ M ammonium sulphate gradient. On the same adsorbent (21.5 $\times$ 150 mm bed) it is also possible to purify crude samples of lipoxidase (200 mg), phosphoglucose isomerase (100 mg) and L-lactate dehydrogenase (54 mg) in less than 100 min, 60 min and 120 min, respectively (20).

7.7 Resolution of small molecular weight biomolecules

Whilst HPAC has found wide application in the resolution and purification of large biomolecules such as proteins and enzymes, it should not be forgotten that it also finds application as 'reversed affinity' HPAC technique to resolve small molecular weight biomolecules. To illustrate this statement a short-description of three such separations will now be given.

7.7.1 Application to adenine nucleotides

When a mixture containing AMP (0.5 nmol), ADP (0.2 nmol) and ADP-ribose (1 nmol) is applied to an alcohol dehydrogenase-silica column [see Section 4.2.3(iii)] under isocratic conditions, separation of the three species is achieved on the basis of the different strength of interaction between the silica-immobilized enzyme and the three nucleotides (*Figure 6*) (21).

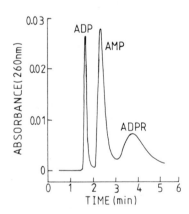

Figure 6. Resolution of adenine nucleotides on alcohol dehydrogenase-silica. HPAC column, alcohol dehydrogenase-silica (12 mg/g), 5 cm $\times$ 0.5 cm; sample applied, 0.5 nmol AMP, 0.2 nmol ADP and 1 nmol ADP-ribose; flow-rate, 1 ml/min; solvent system, 0.25 M sodium phosphate buffer (pH 7.5) containing 1 μM ZnSO$_4$. Nucleotides were detected by following the absorbance at 260 nm. Reproduced from (19) by permission.

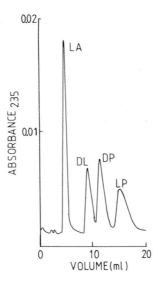

Figure 7. Resolution of D and L amino acid derivatives on bovine serum albumin-silica. HPAC column, bovine serum albumin-silica (Macherey and Nagel & Co., FRG) 15 cm × 0.46 cm; sample applied, 10 μl containing *N*-benzoyl-D,L-alanine; DA and LA (83 μM) and *N*-benzoyl-D,L-phenylalanine; DP and LP (98 μM); flow-rate, 1 ml/min; solvent system, 0.05 M sodium phosphate buffer (pH 5.7). Amino acids were detected by following the absorbance at 235 nm. Reproduced from (20) by permission.

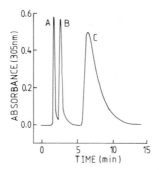

Figure 8. Resolution of glucosides on concanavalin A-silica. HPAC column, silica-bound concanavalin A (25 mg/ml, 10 μm, 10^2 nm), 10 cm × 0.5 cm; sample applied, 3 ml containing 1 μg *p*-nitrophenyl-β-D-glucoside (**A**), 1.5 μg *p*-nitrophenyl-α-D-glucoside (**B**) and 8 μg *p*-nitrophenyl-α-D-mannoside (**C**); flow-rate, 1 ml/min; solvent system, 0.025 M Tris−HCl buffer (pH 6.8) containing 0.25 M NaCl, 0.5 mM $CaCl_2$ and 0.5 mM $MnCl_2$. Sugar derivatives were detected by following the absorbance at 305 nm. Reproduced from (22) by permission.

7.7.2 *Application to the optical isomers of amino acid derivatives*

Earlier studies in free solution have demonstrated enantioselective properties of albumins. Recently, this phenomenon has been exploited in the resolution of optical isomers of various amino acid derivatives on silica-immobilized bovine serum albumin. Thus a mixture (10 μl) of *N*-benzoyl-D,L-alanine (83 μM) and *N*-benzoyl-D,L-phenylalanine (98 μM) can be resolved into the four isomers on a BSA-silica column under isocratic conditions (*Figure 7*) (22). This column is now commercially available from Macherey-Nagel GmbH, FRG.

7.7.3 *Application to carbohydrates (glucosides)*

Concanavalin A (Con A) is a well known affinity protein-ligand for certain carbohydrates (Chapter 8 discusses the use of such support materials in detail). This affinity has been exploited in the resolution of closely related sugars on silica-bound Con A (18). *Figure 8* demonstrates that a mixture of the three sugars p-nitrophenyl-β-D-glucoside, p-nitro-phenyl-α-D-glucoside and p-nitrophenyl-α-D-mannoside, is resolved into three distinctive peaks in less than 10 min on a Con A-silica column.

7.8 **Large scale HPAC: application to rabbit muscle L-lactate dehydrogenase**

Affinity chromatography has found widespread application in the purification technology of enzymes but only limited use, so far, at a preparative scale. Recently and for the first time HPAC employing a silica-immobilized dye has been successfully engaged in enzyme purifications at the process scale (23). A sample of lactate dehydrogenase from crude rabbit muscle extract containing 1.8 g of protein (61 740 units of enzyme) was loaded on a 3.3 litre column containing silica-Procion blue MX-R. The majority of the inert proteins washed through unbound whereas L-lactate dehydrogenase was adsorbed and subsequently eluted with a pulse of its co-factor NADH (7 mM). This 3.3 litre column was operated well below its binding capacity and afforded enzyme purified more than 8 times at 50% overall yield.

8. REFERENCES

1. Clonis,Y.D. and Small,D.A.P. (1987) In *Reactive Dyes in Protein and Enzyme Technology*. Clonis,Y.D., Atkinson,A., Bruton,C. and Lowe,C.R. (eds), McMillan Publishing Company, Basingstoke, UK, Chapter 5.
2. Unger,K. (ed.) (1979) *Porous Silica*. Elsevier, Amsterdam.
3. Regnier,F.E. and Noel,R. (1976) *J. Chromatogr. Sci.*, **14**, 316.
4. Larsson,P.-O. (1984) In *Methods in Enzymology*. Jakoby,W.B. (ed.), Academic Press, New York, vol. 104, p. 212.
5. Stout,R.W. and Destefano,J.J. (1985) *J. Chromatogr.*, **326**, 63.
6. Dean,P.D.G., Johnson,W.S. and Middle,F.A. (eds) f(1985) *Affinity Chromatography: A Practical Approach*. IRL Press, Oxford, UK.
7. Hjerten,S. (1984) *Trends Analyt. Chem.*, **3**, 87.
8. Kinkel,J.N., Anspach,B., Unger,K.K., Wieser,R. and Brunner,G. (1984) *J. Chromatogr.*, **297**, 167.
9. Turková,J., Bláha,K., Horácek,J., Vojcner,J., Frydrychová,A. and Coupek,J. (1981) *J. Chromatogr.*, **215**, 165.
10. Anderson,D.J. and Walters,R.R. (1985) *J. Chromatogr.*, **331**, 1.
11. Shimura,K., Kazama,M. and Kasai,K.-I. (1984) *J. Chromatogr.*, **292**, 369.
12. Greenwald,R., Chaykovsky,M. and Corey,E.J. (1962) *J. Org. Chem.*, **28**, 1128.
13. Fischer,E. (1904) *Chem. Ber.*, **37**, 2486.
14. Clonis,Y.D. (1987) *J. Chromatogr.*, **407**, 179.
15. Lowe,C.R., Glad,M., Larsson,P.-O., Ohlson,S., Small,D.A.P., Atkinson,A. and Mosbach,K. (1981) *J. Chromatogr.*, **215**, 303.
16. Taylor,R.F. and Marenchic,I. (1984) *J. Chromatogr.*, **317**, 193.
17. Ohlson,S., Hansson,L., Larsson,P.-O. and Mosbach,K. (1978) *FEBS Lett.*, **93**, 5.
18. Guildford,H., Larsson,P.-O. and Mosbach,K. (1972) *Chem. Scripta*, **2**, 165.
19. Kato,Y., Nakamura,K. and Hashimoto,T. (1986) *J. Chromatogr.*, **354**, 511.
20. Kato,Y., Kitamura,T. and Hashimoto,T. (1985) *J. Chromatogr.*, **333**, 202.
21. Nilsson,K. and Larsson,P.-O. (1983) *Analyt. Biochem.*, **134**, 60.
22. Allenmark,S., Bomgren,B. and Boren,H. (1983) *J. Chromatogr.*, **264**, 63.
23. Clonis,Y.D., Jones,K. and Lowe,C.R. (1986) *J. Chromatogr.*, **363**, 31.

CHAPTER 7

HPLC of oligonucleotides

ALFRED PINGOUD, ANJA FLIESS and VERA PINGOUD

1. INTRODUCTION

In the past few years the demand for synthetic oligodeoxynucleotides has increased enormously because of their usefulness in many areas of the life sciences. They are needed:

(i)   for the sequencing of genes;
(ii)   as linkers or adapters for recombinant DNA work;
(iii)   as probes for the isolation of specific genes;
(iv)   as mismatch primers to introduce site-specific mutations into DNA;
(v)   as substrates for enzymological or structural studies; and
(vi)   as probes for the detection of specific DNA sequences in the diagnosis of genetic diseases, of various chronic infections or for forensic purposes.

The fact that oligodeoxynucleotides have become readily available over the last few years is due mainly to the progress made in the solid phase synthesis of DNA which nowadays can be carried out conveniently on manually controlled or fully automatic DNA synthesizers. The synthesis of oligodeoxynucleotides would not be as successful as it is, however, if there were not efficient HPLC procedures available for both the analysis of the crude product mixtures resulting from oligodeoxynucleotide syntheses and the preparative purification of the desired products from these mixtures. It is not a coincidence, therefore, that most of the protocols used for HPLC of oligodeoxynucleotides were introduced by chemists engaged in oligodeoxynucleotide synthesis. Their protocols were subsequently taken up by biochemists and molecular biologists and modified to their particular needs, for example, for the isolation of naturally occurring oligo- and polynucleotides (tRNAs, rRNAs, viroid and viral RNAs, fragments thereof, as well as viral and plasmid DNAs, restriction fragments, etc.), or to monitor enzymatic and chemical reactions involving oligonucleotides (phosphorylation, ligation, cleavage, chemical modification).

We will introduce in this chapter first the chromatographic procedures mainly used for the HPLC of oligonucleotides and discuss their underlying principles. We will then discuss practical aspects and finally present specific examples which will cover most of the principal applications of HPLC of oligo- and polynucleotides.

2. PRINCIPLES AND TECHNIQUES

Oligonucleotides are characterized by various features which allow them to be separated

183

by different types of chromatography.

(i)   They are polyanions which may differ in their charge and, therefore, can be subjected to ion-exchange chromatography.

(ii)   They are made up of lipophilic nucleobases, whose composition and sequence will determine their chromatographic behaviour on reversed-phase columns.

(iii)   Since, furthermore, quite often oligonucleotides to be separated differ by their chain length, size-exclusion chromatography can also be useful for their analysis or purification.

The principles of these three types of chromatography are the same in normal, that is low-pressure liquid chromatography, and in HPLC. It involves the partitioning of solutes between a mobile phase—the eluant, and the stationary phase—the chromatographic support with its particular surface. It is the nature of the chromatographic support, virtually incompressible and in most cases spherical particles of defined diameter and pore size that distinguishes HPLC from normal liquid chromatography and determines its high resolution, superior sensitivity and speed of operation.

2.1 Anion-exchange chromatography

Anion-exchange chromatography has been used for the analysis and purification of nucleic acids ever since nucleic acids and their constituents were separated by chromatography. In anion-exchange chromatography oligonucleotides are bound to the positively charged groups of the chromatographic support and displaced during the chromatographic separation by the negative ions of an increasing salt gradient. In general, the more highly charged the oligonucleotide is, the more firmly it is bound to the ion-exchange resin and the higher the salt concentration has to be to elute the oligonucleotide. Since the number of anionic charges of an oligonucleotide is proportional to its chain length, oligonucleotides are eluted according to their length. Exceptions from this rule are observed when interactions other than electrostatic ones are present between solute and stationary phase or when solute molecules interact with each other, for example with oligonucleotides that form double strands. These undesired interactions can be minimized by proper selection of the column material and the eluant. Ideally, anion-exchange HPLC should separate oligonucleotides strictly according to chain length, such that one can predict the elution volume or retention time of a given oligonucleotide on the basis of the chromatographic behaviour of an oligonucleotide standard consisting of a few oligonucleotides of known chain length.

A variety of anion-exchange HPLC column packing materials have been developed that are suitable for the fractionation of oligonucleotides according to chain length. They consist of alkylamine-derivatized or polyethyleneimine-coated silica particles as well as rigid organic polymers carrying basic functional groups. A selection of commercially available anion-exchange HPLC column packing materials generally used for the separation of oligonucleotides is listed in *Table 1*. Probably the most widely used anion-exchange resin employed for HPLC of synthetic oligodeoxynucleotides is Partisil strong anion-exchanger (SAX) (1). Chromatography on SAX columns is usually carried out with a gradient of potassium phosphate at near neutral pH in the presence of formamide. The addition of formamide to the eluant is beneficial for the resolution in general, since it suppresses non-electrostatic interactions between oligonucleotides and the support.

Table 1. Anion-exchange HPLC column packing materials.

Name	Support	Functional group	Particle diameter (μm)	Pore size (nm)
Partisil SAX (Whatman)	Silica	Quaternary amino (bonded)	10	10
Mono Q (Pharmacia)	Hydrophilic acrylic polymer	Quaternary amino	10	70
Nucleogen 60	Silica	Tertiary amino (bonded)	7	6
Nucleogen 500			10	50
Nucleogen 4000 (Macherey & Nagel)			10	400
TSK DEAE 3 SW (Toyo Soda)	Silica	Tertiary amino (bonded)	10	25
TSK DEAE 5 PW (Toyo Soda)	Hydroxylated polyether	Tertiary amino (bonded)	10	100
PEI widepore (Baker)	Silica	Primary and secondary amino (pellicular)	5	33
LiChrosorb NH$_2$ (Merck)	Silica	Primary amino (bonded)	5	6

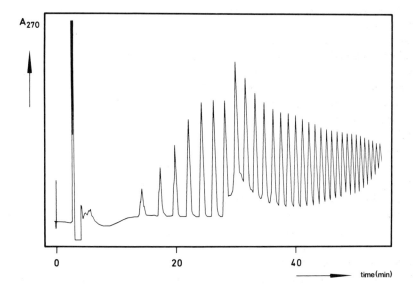

Figure 1. Analytical anion-exchange HPLC of a commercial oligo(dA) preparation on a SAX column. Sample, 10 μg oligo(dA); column, Whatman Partisil 10 SAX (4.6 × 250 mm); apparatus, DuPont 850 liquid chromatograph with a DuPont UV spectrophotometer set at 270 nm with 0.04 AUFS (absorbance units full scale); elution, linear gradient from 1 mM potassium phosphate, pH 6.3, 60% (v/v) formamide to 0.3 M potassium phosphate, pH 6.3, 60% (v/v) formamide in 75 min; flow-rate, 1 ml/min.

For sequences that are partially self-complementary or which have high proportions of guanine residues or more than two consecutive guanine residues eluants containing 60% (v/v) formamide are recommended (2). For the sake of convenience, we use SAX columns only with buffers containing 60% (v/v) formamide, irrespective of the composition or sequence of the oligonucleotides to be separated. A chromatogram which

185

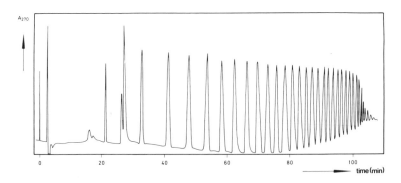

Figure 2. Analytical anion-exchange HPLC of an oligo(rA) preparation on a PEI column. Sample, 20 μg commercial poly(rA) digested with 1 N NaOH at 56°C for 3 min and then neutralized with 1 N HCl; column, Baker PEI wide pore (4.6 × 250 mm); apparatus, Bruker LC 21 B with a Shimadzu SPD-6A spectrophotometer set at 270 nm with 0.04 AUFS; elution, linear gradient from 1 mM potassium phosphate, pH 6.3, 60% (v/v) formamide to 0.225 M potassium phosphate, pH 6.3, 60% (v/v) formamide in180 min; flow-rate, 1 ml/min.

demonstrates the resolution that can be achieved with anion-exchange HPLC of oligonucleotides on SAX columns is shown in *Figure 1*. It can be estimated that oligonucleotides with a chain length up to 30 can be resolved. The retention times of oligonucleotides have been shown to be almost independent of base composition or sequence under optimum conditions (3). HPLC on SAX columns, however, suffers from a major disadvantage. Even when properly treated (*vide infra*) the SAX column packing material is not very stable in aqueous media and columns deteriorate quickly.

Other anion-exchange resins have been developed which do not suffer from this drawback. We routinely now use polyethyleneimine (PEI) columns for the separation of oligonucleotides. PEI columns can be prepared by coating silica particles *in situ* with polyethyleneimine (4). The original protocol recommended an ammonium sulphate gradient in phosphate buffer in the presence of methanol for the separation of oligonucleotides. We run commercially available PEI wide pore columns with the same buffer system as described for the SAX columns; that is solvent A, 1 mM potassium phosphate, pH 6.3, 60% (v/v) formamide; and solvent B, 0.3 M potassium phosphate, pH 6.3, 60% (v/v) formamide.

PEI columns have a similar resolution as SAX columns: oligonucleotides with a chain length up to 30 can be separated (*Figure 2*). It was reported recently that quaternization of the PEI anion-exchange material by methylation with methyl iodide improves the resolving power of PEI columns, such that oligonucleotides up to a chain length of 50 can be resolved (5). This column packing material, however, is not commercially available. PEI wide pore columns have a high capacity such that more than 10 mg of oligonucleotides can be separated in one run on an analytical column (4.6 mm × 250 mm) (*Figure 3*).

Quite similar in performance as the PEI is the Nucleogen DEAE column packing material which is available with different pore sizes, the 6 nm material being recommended for the semi-preparative separation of oligodeoxynucleotides, while the 50 and 400 nm materials have been developed for the high-resolution separation of

186

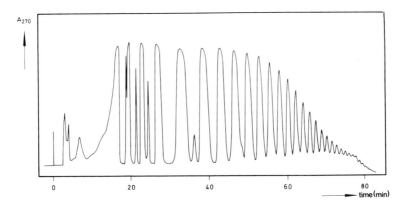

Figure 3. Preparative anion-exchange HPLC of an oligo(rA) preparation on a PEI column. Sample, 5.6 mg oligo(rA); column, Baker PEI wide pore (4.6 × 250 mm); apparatus, Bruker LC 21 B with a Shimadzu SPD-6A spectrophotometer set at 270 nm with 2.56 AUFS; elution, linear gradient from 1 mM potassium phosphate pH 6.3, 60% (v/v) formamide to 0.3 M potassium phosphate, pH 6.3, 60% (v/v) formamide in 180 min. Flow-rate, 1 ml/min. It is important to note that the Baker PEI wide-pore column used here does not have memory effects after such preparative runs; this means that in subsequent analytical runs no peaks due to the previous preparative run appear. With the Nucleogen DEAE 60-7 column which we have also tested for preparative separations of oligodeoxynucleotides slight memory effects are apparent.

oligonucleotides as well as the analysis and purification of naturally occuring high-molecular-weight nucleic acids (6).

TSK DEAE 3 SW and TSK DEAE 5 PW column materials (7) are reported to resolve oligonucleotides with a chain length up to 30. We have no experience with these two column packing materials and therefore cannot critically compare their performance with those of others. According to a recent publication, however, the TSK DEAE 5 PW column does not perform as well as the Nucleogen DEAE 4000 or the Mono Q columns in the separation of DNA restriction fragments (8).

Recently the strong anion-exchange resin Mono Q, so far used mainly for fast protein liquid chromatography, was introduced as a column material suitable also for the purification of synthetic oligodeoxynucleotides (9) as well as for the separation of restriction fragments (8). Since this column packing material is based on an organic polymer rather than on silica, it can be used in a wide pH range (pH 2−12). At alkaline pH the Mono Q column very efficiently separates oligodeoxynucleotides strictly according to chain length, and performs better in this respect than the Partisil SAX column at neutral pH (9).

Quite a few aminoalkyl derivatized silica column packing materials are commercially available (e.g. Zorbax-NH_2 from DuPont, LiChrosorb-NH_2 from Merck). They can be used as anion-exchange resins for oligonucleotides and compare favourably with the Partisil SAX column material with respect to resolution and, in particular, chemical stability (10). Nevertheless they were never widely used for the separation of oligonucleotides, presumably because they do not fractionate oligonucleotides strictly according to chain length. This feature, however, can be used advantageously, for example for the separation of sequence isomers of oligodeoxynucleotides or for the fractionation of tRNA fragments (Section 4.8) which—when larger—cannot be resolved by reversed-phase chromatography.

187

Table 2. Reversed-phase HPLC column packing materials.

Name	Functional group	Particle diameter (μm)	Pore size (nm)
μBondapack C$_{18}$ (Waters)	C$_{18}$	10	10
Zorbax ODS	C$_{18}$	3,7	10
C$_8$	C$_8$	3,7	10
PEP-RP1 (DuPont)	C$_8$	4	30
Partisil ODS	C$_{18}$	5,10	
CCS	C$_8$	5,10	
Protesil Octyl (Whatman)	C$_8$	10	20
Nucleosil C$_4$	C$_4$	5,7,10	30
C$_8$	C$_8$	3,5,7,10	5,10,12,30
C$_{18}$ (Macherey & Nagel)	C$_{18}$	3,5,7,10	5,10,12,30
LiChrosorb RP8	C$_8$	5,7,10	6
RP18 (Merck)	C$_{18}$	5,7,10	6
Bakerbond C$_8$	C$_8$	5	5
C$_{18}$	C$_{18}$	5	5
Baker Widepore C$_4$	C$_4$	5	33
C$_8$	C$_8$	5	33
C$_{18}$ (Baker)	C$_{18}$	5	33
Biosil ODS (Bio-Rad)	C$_{18}$	5,10	10

2.2 Reversed-phase chromatography and reversed-phase ion-pair chromatography

In reversed-phase chromatography of oligonucleotides advantage is taken of the lipophilic character of nucleic acids themselves or—as in reversed-phase ion-pair chromatography—of the alkylammonium salts of nucleic acids (11,12). Oligonucleotides are bound to the non-polar column packing material by hydrophobic interactions, elution is achieved normally by addition of organic solvents, most often acetonitrile, to the eluant. These hydrophobic interactions increase with chain length and are more pronounced with single-stranded oligonucleotides than with double-stranded ones. The effects of base composition and sequence are considerable, such that separation is not strictly according to size (3). As a matter of fact, it is precisely this feature of reversed-phase HPLC that makes this type of chromatography an ideal companion to anion-exchange HPLC. Since the chromatographic behaviour of a given oligonucleotide on reversed-phase columns is not easy to predict, reversed-phase HPLC cannot be recommended for the identification of an oligodeoxynucleotide as the desired product of a chemical synthesis, but is very useful for an estimation of its purity.

The classical procedure for reversed-phase HPLC of oligonucleotides involves chromatography on octadecylsilyl columns in ammonium or triethylammonium acetate buffer with an acetonitrile gradient (11).

A variety of different reversed-phase column packing materials have been used for the separation of oligonucleotides. *Table 2* gives a selection of commercially available column packing materials. It includes C_4, C_8 and C_{18} modified silica particles. In general, the retention of oligonucleotides follows the order C_{18}, C_8, C_4 but depends also on the total carbon loading of the column packing material. We generally use C_{18} columns for reversed-phase HPLC of oligonucleotides and C_8 or C_4 columns for reversed-phase ion-pair HPLC of oligonucleotides. For highest resolution column packing materials with particles of small diameter and large pore size should be used. When high capacity is needed particles with small pore size are recommended.

The nature of the counter-ion present has a considerable effect on the chromatographic process in reversed-phase HPLC of oligonucleotides. For example, with ammonium acetate buffer oligonucleotides are eluted earlier from the column than with triethyl-ammonium acetate buffer (11). With ammonium ions as counter-ions the retention of an oligonucleotide on a reversed-phase column is a function mainly of its hydrophobicity, while with triethylammonium acetate ions as counter-ions ion-pairing phenomena come into play. This means that the oligonucleotides are adsorbed to the stationary phase not directly but rather via their counter-ions. The strength of interaction of the oligonucleotide with the reversed-phase support is then determined by the hydrophobicity of the counter-ion. Accordingly in the presence of tetrabutylammonium hydrogen sulphate (or phosphate) which is very hydrophobic and, therefore, can be used as an ion-pairing reagent, oligonucleotides are bound more firmly to the reversed-phase column than in the presence of triethylammonium or ammonium acetate. Furthermore, since the strength of interaction increases with the charge of the oligonucleotide, reversed-phase columns equilibrated with tetrabutylammonium hydrogen sulphate function in essence as anion-exchange columns for oligonucleotides. In such a reversed-phase ion-pair chromatographic mode, oligonucleotides are separated predominantly according to chain length (3) in contrast to the separation achieved using normal reversed-phase chromatography. The resolution of oligonucleotides, however, is limited to a chain length of approximately 15 (*Figure 4*). Since reversed-phase column packing materials are chemically very stable and comparatively cheap, reversed-phase ion-pair chromatography is the method of choice for the separation of short oligonucleotides according to chain length. As a matter of fact with one reversed-phase column and with two different buffer systems, namely:

(I) solvent A: 0.1 M triethylammonium acetate, pH 7.0, 1% acetonitrile,
     solvent B: 0.1 M triethylammonium acetate, pH 7.0, 50% acetonitrile;
(II) solvent A: 50 mM potassium phosphate, pH 5.9, 2 mM tetrabutylammonium hydrogen sulphate,
     solvent B: 50 mM potassium phosphate, pH 5.9, 2 mM tetrabutylammonium hydrogen sulphate, 60% acetonitrile,

one can analyse short oligonucleotides both with respect to homogeneity using system I and identity (chain length) using system II.

2.3 Mixed mode chromatography

Mixed mode chromatography of oligonucleotides takes advantage of both hydrophobic and electrostatic interactions between the oligonucleotides to be separated and the

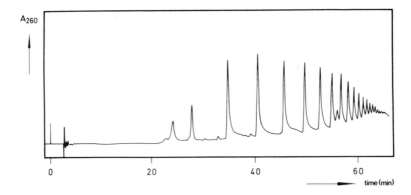

Figure 4. Analytical reversed-phase ion-pair HPLC of oligo(rA) on a C_{18} column. Sample, 10 µg oligo(rA); column, LiChrosorb RP-18 (4.0 × 250 mm); apparatus, Bruker LC 21 B with a Shimadzu SPD-6A spectrophotometer set at 260 nm with 0.16 AUFS; elution, linear gradient from 0 to 60% (v/v) acetonitrile in 50 mM potassium phosphate, pH 5.5, 2 mM tetrabutylammonium hydrogen sulphate in 120 min; flow-rate, 0.8 ml/min.

stationary phase. The classical example of a mixed mode column packing material is the low-pressure resin RPC-5 (13). It consists of non-porous Plaskon or Teflon particles to which aliphatic quaternary ammonium salts are adsorbed. RPC-5 has been applied successfully to the separation of tRNAs, tRNA fragments and more recently to the isolation of restriction fragments (14). RPC-5 itself is not commercially available. A RPC-5 analogue, however, has been developed, NACS, which can be purchased from BRL. NACS-12 has been used for mixed mode HPLC of DNA, DNA fragments and synthetic oligodeoxynucleotides.

McLaughlin and colleagues have derivatized aminopropyl silica particles with alkyl and aryl residues in order to obtain hydrophobic anion-exchange supports suitable for mixed mode chromatography. These supports were successfully used for separation of sequence isomeric oligodeoxynucleotides and tRNA fragments (15).

Hartwick and co-workers have bonded alkyl- and chloroalkylsilane to silica particles and subsequently substituted chlorine by a tertiary amine. This mixed mode support was used for the separation of crude oligodeoxynucleotides and restriction fragments (16).

Silica-based mixed mode column packing materials are not yet commercially available. At present it seems as if mixed mode chromatography is useful only for very specialized problems, for example separation of sequence isomeric oligonucleotides.

2.4 Size-exclusion chromatography

In size-exclusion or gel-permeation chromatography of oligo- and polynucleotides separation is according to size. Several HPLC column packing materials suitable for this purpose have been developed. They consist of organic polymers or silica particles with hydrophilic groups or coated with hydrophilic polymers. They all have pores of defined diameter in order to achieve separation by size: large oligo- or polynucleotides will be excluded from these pores and will be eluted with the void volume, while small oligonucleotides will be able to diffuse into these pores and will be eluted later. Size-exclusion HPLC has been used for the separation of oligodeoxynucleotide (17), ribosomal RNAs and restriction fragments (18). At present anion-exchange HPLC seems to be superior to size-exclusion HPLC with respect to resolution in particular in the separation of oligonucleotides, not as much as in the fractionation of high-molecular-weight nucleic acids.

3. PRACTICAL CONSIDERATIONS

In the following paragraph we will at first discuss what kind of equipment is needed for HPLC of oligodeoxynucleotides. We will then consider practical aspects concerning the handling of HPLC columns and finally we will give information as to the preparation of solvents and samples.

3.1 **Apparatus**

Commercial high-performance liquid chromatographs usually comprise one or more high-pressure pumps, (depending on whether the gradient is mixed under low or high pressure), a gradient controller, a UV spectrophotometer with a recorder and a manually operated injector. Such basic systems, like the Waters 600 or the Merck-Hitachi 655 chromatographs, are sufficient for most applications. As a matter of fact equipment of this kind has been used for all but one application detailed in Section 4. Two methods of gradient formation are employed.

(i)   The gradient is mixed on the high-pressure side of the pump, in general in conjunction with a separate pump for each solvent. The advantage of this arrangement is that gas bubbles formed upon solvent mixing will not disturb solvent delivery. A disadvantage of this system is that gradients are not very accurate when low amounts of solvent A or B have to be delivered by the pumps, usually at the beginning or the end of a gradient.

(ii)   The gradient is mixed on the low-pressure side of the pump; two or more electromagnetic proportioning valves direct the flow from the different solvent reservoirs to the mixing chamber and the pump. The main advantage of this arrangement is that only one pump is needed and that gradient formation is very accurate. A disadvantage of this system is the greater sensitivity of the solvent delivery to gas bubble formation. It seems to us, however, that at least with modern pumps this is a minor point, since it can be easily overcome by de-gassing the solvent.

Whatever configuration is chosen, it should allow for an eluant flow of at least up to 10 ml/min, in order to cope with preparative separations.

There are some accessory pieces of equipment available which although not absolutely required make analytical and preparative work with oligonucleotides much easier.

(i)   A thermostatable column compartment ('column oven') will allow chromato-graphy of oligonucleotides at elevated temperatures which results quite often in enhanced resolution in particular with self-complementary oligonucleotides.

(ii)   A variable UV spectrophotometer is to be given preference over a fixed-wavelength detector because it allows de-tuning of the instrument in order to reduce the detector sensitivity for preparative separations and thereby to circumvent the installment of preparative flow-through sample cells.

(iii) An integrating recorder is not necessary for most purposes presented in Section 4. As soon, however, as quantitative aspects dominate in HPLC of oligonucleotides, it is a must.

3.2 **Column**

The different kind of column packing materials generally employed for HPLC of oligonucleotides have been discussed in Section 2. All of these column packing materials can be purchased in pre-packed columns, most of them are also available in bulk for the self-packing of columns. Since HPLC columns have only a very limited life-time, even when adequately handled, column costs are substantial in HPLC of oligonucleotides. If HPLC, therefore, is a routine procedure within one's work and financial resources are rather limited, one should seriously consider the use of self-packed columns. Reliable and straightforward procedures for the packing of columns have been published (19). Self-packed columns when properly prepared, are indistinguishable in performance from commercial pre-packed columns.

The life-time of a column is very much dependent on how the column is operated. This is particularly true for the very delicate anion-exchange columns. The following recommendations for the operation of columns when followed are beneficial for the life time of HPLC columns.

(i) Columns should always be used together with a guard column which are placed between the injection port and the column. Care, however, should be taken to minimize the length of the tubing connecting the guard column, the injection port and the column. The guard column retains material which is irreversibly adsorbed to the column packing material and, thereby, will protect the main column from accumulating such material. Guard columns are replaced (emptied and re-filled), when the back-pressure of the system gets too high or the resolution is impaired compared to previous runs. Guard columns can be purchased pre-packed at low cost; since self-packing of guard columns using a methanol slurry of the column packing material is very easy, we recommend, the use of self-packed guard columns.

(ii) Even when guard columns are used and chromatography is carried out only with filtered eluants and samples (*vide infra*), the back-pressure of the column itself will increase eventually. This does not necessarily mean that the column must be replaced. Very often a cleaning of the inlet frit with concentrated nitric acid, preferably under ultrasonic agitation, and regeneration of the column packing material by washing the column first with methanol and then with a methanol/chloroform 1:1 (v/v) mixture will restore the resolution of the column.

(iii) Columns containing silica-based packing materials should only be used with buffers having a pH between 2.5 and 8.0, since the silica matrix is sensitive to alkaline pH. Columns should not be subjected to unnecessary high salt concentrations or high pH, not even for storage overnight. We recommend that columns should be stored in water for short-term storage (up to 24 h) and in methanol for longer periods of time. Since most columns are made of stainless steel, corrosive chemicals should be avoided and these include halide ions and

EDTA which, therefore, should only be used when necessary for the resolution. Anion-exchange column packing materials are particularly sensitive to destruction, because their functional groups are sensitive to oxidation and other electrophilic reactions.

(iv)   The long-term performance of a column is affected by mechanical stresses which arise from abrupt changes of flow-rate or solvent composition. It is therefore advisable to increase the flow-rate stepwise until the desired flow-rate is reached, as well as to avoid an abrupt change in solvent composition by using a steep gradient for example when returning to starting conditions after the end of a gradient elution.

(v)   If it can be afforded, a column should be used for a particular application only thereby avoiding changes between different solvent systems.

3.3 Eluants

Solvents and buffer salts used for HPLC should be of analytical grade and thus as much as possible devoid of contaminants since these latter tend to get adsorbed onto the column, either irreversibly or reversibly. When this occurs, the contaminants are often eluted from the column in the course of a gradient-elution and produce 'ghost peaks'. Many organic solvents commonly employed for HPLC are commerically available as HPLC grade. *Table 3* gives a compilation of various solvents and buffer salts which are needed for HPLC or oligonucleotides.

In order to avoid blocking of the column by small particles all eluants should be filtered prior to use. For this purpose we employ Whatman GF/F glass microfibre filters. Eluants should also be thoroughly de-gassed, in order to avoid the formation of gas bubbles in the mixing chamber and at the column exit which will interfere with solvent delivery and detection. Eluants can be de-gassed either by stirring for 5 min *in vacuo* using a water pump and a magnetic stirrer or by flushing with helium.

Table 3. Solvents and other chemicals used for HPLC of oligonucleotides.

Chemical	Specification	Source	Catalogue number
Acetonitrile	HPLC gradient grade	Baker	8143
Formamide	p.a.	Fluka	47670
Methanol	For HPLC	Baker	8402
Water	Quartz bidistilled	–	–
KH_2PO_4	p.a.	Merck	4873
NaH_2PO_4	p.a.	Merck	6346
NaCl	p.a.	Baker	0278
Urea	p.a.	Merck	8487
Tetrabutylammonium hydrogen sulphate	puriss.	Fluka	86868
Triethylammonium acetate			
Triethylamine	p.a.	Fluka	90340
Acetic acid	p.a.	Riedel-de Haen	33209

p.a., pro analysi; puriss., purissimum.

3.4 **Sample preparation and work-up**

The sample should be dissolved in starting buffer or if already dissolved, adjusted as much as possible to starting conditions. This is of particular importance for preparative HPLC in which very often large sample volumes have to be injected into the column. The sample to be loaded onto the column should be free of particulate matter. This can be achieved either by filtration or by centrifugation.

Recovery of purified oligonucleotides after preparative HPLC is straightforward when volatile eluants are used as in most reversed-phase separations, since in this case lyophilization or preferably evaporation *in vacuo* using a Speed-Vac concentrator (Savant) will both concentrate and de-salt the sample. When eluants containing non-volatile constitutents are used, as in most anion-exchange HPLC protocols, recovery of the purified oligo- or polynucleotides can be achieved by four methods.

Method 1

(i)   Dilute the combined fraction with water to an ionic strength of 0.2 M.
(ii)   Load onto a DE 52 (Whatman) column (0.4 × 2 cm) equilibrated with 0.3 M triethylammonium acetate, pH 7.0, buffer.
(iii)   Wash the column with approximately 20 ml of equilibration buffer.
(iv)   Elute the oligonucleotide with 1 M triethylammonium acetate, pH 7.0.
(v)   Lyophilize.
(vi)   Take up in water/ethanol 1:1 (v/v) and lyophilize again.

Method 2

(i)   Dialyse the combined fractions extensively against water.
(ii)   Lyophilize.
(iii)   Take up in water/ethanol 1:1 (v/v) and lyophilize again.

Method 3

Polynucleotides, for example restriction fragments, can be concentrated from dilute solution (down to 10 ng/ml) by precipitation with 10% (w/v) polyethyleneglycol.

(i)   Add solid polyethyleneglycol (PEG 6000) to the fractions of interest.
(ii)   Leave on ice for 2 h.
(iii)   Centrifuge.
(iv)   Wash the pellet with 75% (v/v) ethanol/water.

Method 4

Alternatively, precipitaton can be carried out with isopropanol, when the combined polynucleotide fractions contain less than 1 M NaCl or KCl.

(i)   Add an equal volume of isopropanol.
(ii)   Leave on ice for 1 h.
(iii)   Centrifuge.
(iv)   Wash the pellet with 75% (v/v) ethanol/water.

Method 1 is the method of choice for oligonucleotides with a chain length below 10. Method 2 is somewhat time consuming due to the dialysis step, but is not labour-intensive and gives the best yields of all four methods. Methods 3 and 4 are applicable with good yield only for oligonucleotides with a chain length above 20.

4. APPLICATIONS

Several applications of HPLC of oligo- and polynucleotides will be described in this section. The examples which we have selected represent procedures of proven performance and reliability. In several instances they are presented together with detailed protocols.

4.1 Purification and analysis of oligodeoxynucleotides synthesized on solid phase supports

Purification of oligodeoxynucleotides synthesized on solid phase supports is necessary for most purposes since these oligodeoxynucleotides are contaminated by shorter oligodeoxynucleotides, due to incomplete reaction within each cycle by oligodeoxynucleotides still carrying protecting groups due to incomplete deprotection at the end of the synthesis. Contamination can also occur by oligodeoxynucleotides that have undergone side reactions, in particular de-purination due to the acid treatment for the removal of the 5'-protecting group at the end of each cycle. Since in solid phase synthesis intermediates are not purified during chain assembly, erroneous products accumulate and must be removed after synthesis. This can be achieved conveniently and effectively by HPLC or by polyacrylamide gel electrophoresis under denaturing conditions. HPLC is to be preferred over polyacrylamide gel electrophoresis, when mg quantities of a synthetic oligodeoxynucleotide are to be purified. Depending on the quality of the crude oligodeoxynucleotide preparation and the purity wanted, one or two HPLC runs have to be performed. If a one-step purification scheme seems to be sufficient, either anion-exchange HPLC (or reversed-phase ion-pair HPLC) of the fully de-protected oligodeoxynucleotide or reversed-phase HPLC of the 5'-DMT (dimethoxytrityl)-oligodeoxynucleotide can be carried out in order to remove shorter oligodeoxynucleotides representing the failure sequences accumulated during the synthesis.

In both HPLC procedures the oligodeoxynucleotide of interest will in general be eluted *after* all contaminant oligodeoxynucleotides and therefore can be readily identified. For anion-exchange HPLC this is due to the fact that the desired product is the longest oligodeoxynucleotide and carries the largest number of phosphate groups. For reversed-phase HPLC, this is due to the fact that only the desired product carries the dimethoxytrityl group, when a 'capping' reaction is carried out after each coupling step; after release of the oligodeoxynucleotide from the support and removal of the base and phosphate protection groups by ammonia treatment, the largest oligodeoxynucleotide as a result of the last coupling step still carries the dimethoxytrityl group which will greatly increase the hydrophobicity of this oligodeoxynucleotide.

We recommend the use of reversed-phase HPLC as the first purification step after release of the oligodeoxynucleotide from the support and removal of the protecting groups other than the dimethoxytrityl group. After de-tritylation of the HPLC purified oligodeoxynucleotide the resulting product should be analysed by anion-exchange HPLC. If the purity is not sufficient, a second chromatographic purification step, anion-exchange HPLC or reversed-phase ion-pair HPLC, can be carried out (*Table 4* and *Figure 5a,b*).

Table 4. Purification of oligodeoxynucleotides synthesized on solid support.

1.	Release the oligodeoxynucleotide from the support and remove the base and phosphate protection groups by treatment with 3% (v/v) ammonia for 3 h at 65°C. Remove ammonia by evaporation to dryness *in vacuo*. Dissolve the residue in ethanol/water (1:1) and evaporate the solvent to dryness *in vacuo*; repeat this procedure.
2.	Dissolve the sample in water and load the solution onto a LiChrosorb RP-18 column (25 × 250 mm). Carry out a gradient-elution from 20 to 100% B in 120 min at a flow-rate of 4 ml/min [solvent A: 0.1 M triethylammonium acetate, pH 7.0, 1% (v/v) acetonitrile; solvent B: 0.1 M triethylammonium acetate, pH 7.0, 50% (v/v) acetonitrile]. 5′-DMT-oligodeoxynucleotides are eluted at ~70% B. Combine the fractions of interest. Remove the solvent by evaporation to dryness *in vacuo*.
3.	Remove the dimethoxytrityl group by treatment with 80% (v/v) acetic acid for 30 min at room temperature. Evaporate to dryness *in vacuo*. Dissolve the residue in water, evaporate to dryness *in vacuo*; repeat this procedure. Dissolve the residue in water, extract three times with ether. Evaporate to dryness *in vacuo*, dissolve the residue in water, evaporate to dryness *in vacuo*.
4.	Dissolve the sample in water. Check the purity by anion-exchange HPLC, for example on a PEI column (4.6 × 250 mm) using a gradient from 0 to 70% B in 60 min at a flow-rate of 0.5 ml/min [solvent A: 1 mM potassium phosphate, pH 6.3, 60% (v/v) formamide; solvent B: 0.3 M potassium phosphate, pH 6.3, 60% (v/v) formamide]. When the purity of the oligodeoxynucleotide is not satisfactory, carry out anion-exchange HPLC on a preparative scale using the same column under the same conditions.

A one-step purification procedure using only reversed-phase HPLC might be critical for longer oligodeoxynucleotides, because often a small percentage of other dimethoxytrityl-bearing oligodeoxynucleotides accumulate in addition to the desired 5′-DMT-oligodeoxynucleotide during solid phase synthesis of oligodeoxynucleotides. They originate most probably from de-purination reactions occuring during the acid treatment used to remove the dimethoxytrityl group at the end of each cycle. The ammonia treatment after completion of the synthesis will then cause phosphodiester bond cleavage at the sites of de-purination leading to a series of shorter 5′-dimethoxytrityl oligodeoxynucleotides. These can be removed by careful fractionation of the peak presenting the dimethoxytrityl-bearing species during preparative reversed-phase HPLC (*Figure 6a−d*). Of particular usefulness for this purpose are wide-pore C_4 columns (20).

4.2 **Separation of phosphorylated and unphosphorylated oligodeoxynucleotides**

Separation of oligodeoxynucleotides phosphorylated at the 5′-position from those having a free 5′-hydroxyl group is necessary for a variety of different research topics, e.g. for structural and mechanistic studies involving phosphorylated oligodeoxynucleotides or for the localization of genes with radioactive phosphorylated oligodeoxynucleotide probes. This separation can be achieved by reversed-phase HPLC as well as by anion-exchange HPLC. *Figure 7* shows the separation of d(GGAATTCC) from its phosphorylated reaction product by anion-exchange HPLC on a SAX column, a separation which was necessary for the study of the influence of the 5′-terminal phosphate group on the kinetics of cleavage of an octadeoxynucleotide substrate by the restriction endonuclease *Eco*RI.

The sensitivity of detection of a specific DNA sequence by hybridization with a

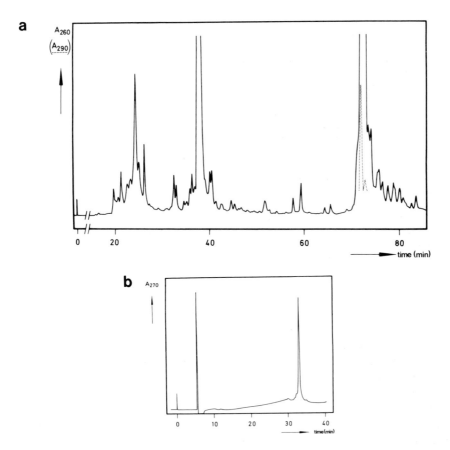

Figure 5. (a) Preparative reversed-phase HPLC of a crude 5'-DMT-decadeoxynucleotide preparation on a C$_{18}$ column. Sample, d(GGGATATGCC) was synthesized on the Omnifit oligodeoxynucleotide synthesizer by the phosphotriester method. The oligodeoxynucleotide was released from the solid support and de-protected by 33% (w/v) NH$_3$ treatment for 3 h at 65°C. Approximately 6 mg of the oligodeoxynucleotide carrying the dimethoxytrityl group on its 5' end was loaded onto the column. Column, LiChrosorb RP-18 (25 × 250 mm); apparatus, Bruker LC 21 B with a Shimadzu spectrophotometer set at 260 or 290 mm, respectively, with 2.56 AUFS; elution, linear gradient from 10 to 50% (v/v) acetonitrile in 0.1 M triethylammonium acetate, pH 7.0, in 120 min; flow-rate, 6 ml/min. Most of the chromatogram was recorded at 260 nm. In the range of interest the spectrophotometer was de-tuned to 290 nm in order to record the total peak height. The oligonucleotide material eluted at 72 min was pooled, fully de-protected and analysed by anion-exchange HPLC. **(b)** Analytical anion-exchange HPLC of a reversed-phase HPLC purified decadeoxynucleotide on a PEI column. Sample, 400 ng of d(GGGATATGCC) purified by reversed-phase HPLC (*Figure 5a*) and relieved of its dimethoxytrityl protection group by incubation with 80% (v/v) acetic acid for 30 min at ambient temperature, lyophilized and extracted three times with ether. Column, Baker PEI wide pore (4.6 × 250 mm); apparatus, Bruker LC 21 B with a Shimadzu SPD-6A spectrophotometer set at 270 nm with 0.08 AUFS; elution, linear gradient from 1 mM potassium phosphate, pH 6.3, 60% (v/v) formamide to 0.21 M potassium phosphate, pH 6.3, 60% (v/v) formamide in 60 min; flow-rate, 0.5 ml/min. The analytical anion-exchange HPLC indicates that the oligodeoxynucleotide preparation after the preparative HPLC step is to over 95% homogeneous.

radiaoctive probe is dependent on the specific activity of the probe. With 5'-end labelled oligodeoxynucleotides the specific activity of a probe is limited by the specific activity of the [γ-^{32}P]ATP itself. It is important, therefore, to separate the labelled from the

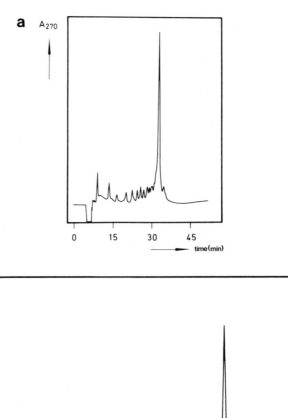

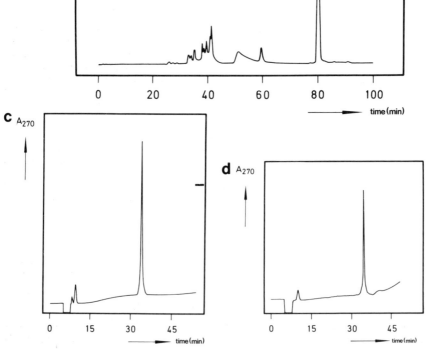

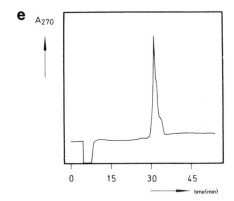

Figure 6. (a) Analytical anion-exchange HPLC of a crude eicosadeoxynucleotide preparation on a PEI column. Sample, d(ACAACTACGCAGCGCCTCCC) was synthesized on a Biosearch DNA synthesizer by the phosphoramidit method, released from the solid support and de-protected by 33% (v/v) NH_3 treatment for 3 h at 65°C. A small part of this sample (~10 μg) was treated with 80% (v/v) acetic acid for 30 min at ambient temperature and subjected to anion-exchange HPLC (this figure) while most of the sample containing the 5'-DMT-oligodeoxynucleotide was subjected to preparative reversed-phase HPLC (b). Column, Baker PEI wide pore (4.6 × 250 mm); apparatus, Merck-Hitachi 655A-12 with a Merck-Hitachi 655A-22 spectrophotometer set at 270 nm with 0.16 AUFS; elution, linear gradient of 1 mM potassium phosphate, pH 6.3, 60% (v/v) formamide to 0.21 M potassium phosphate, pH 6.3, 60% (v/v) formamide in 60 min; flow-rate, 0.5 ml/min. The analytical anion-exchange HPLC of the crude oligodeoxynucleotide preparation was carried out in order to get a rough idea of the quality of the preparation and in order to check the efficiency of the subsequent preparative HPLC purification step (b−e). The chromatogram indicates that the crude reaction mixture contains to ~70% the desired oligodeoxynucleotide. **(b)** Preparative reversed-phase HPLC of a crude 5'-DMT-eicosadeoxynucleotide preparation. Sample, 300 μg 5' -DMT-d(ACAACTA-CGCAGCGCCTCCC) prepared as in a; column, LiChrosorb RP-18 (25 × 250 mm); apparatus, Merck-Hitachi 655A-12 with a Merck-Hitachi 655A-22 spectrophotometer set at 280 nm with 2.56 AUFS; elution, linear gradient of 10−55% (v/v) acetonitrile in 0.1 M triethylammonium acetate pH 7.0, in 130 min; flow-rate, 4 ml/min. Three fractions eluting at 79 min (fraction I), 80 min (fraction II) and 81 min (fraction III) were collected and analysed by anion-exchange HPLC after removal of the dimethoxytrityl protection group (c−e). As is apparent from c−e only fractions I and II contain the desired oligodeoxynucleotide, while fraction III comprises shorter oligodeoxynucleotides, most probably arising from de-purination reactions. **(c−e)** Analytical anion-exchange HPLC of three consecutive fractions of a preparative reversed-phase HPLC of a crude 5'-DMT-eicosadeoxynucleotide preparation. Sample, fractions I, II and III (b) were subjected to a 80% (v/v) acetic acid treatment for 30 min at ambient temperature in order to remove the dimethoxytrityl protection group. The acetic acid was removed by evaporation. Aliquots comprising ~1 μg oligodeoxynucleotide were re-chromatographed by HPLC on a PEI column (this figure); **(c)** fraction I, **(d)** fraction II, **(e)** fraction III. Column, Baker PEI wide pore (4.6 × 250 mm); apparatus, Merck-Hitachi 655A-12 with a Merck-Hitachi 655A-22 spectrophotometer set at 270 nm with 0.04 **(c)**, 0.08 **(d)** and 0.02 **(e)** AUFS; elution, linear gradient from 1 mM potassium phosphate, pH 6.3, 60% (v/v) formamide to 0.21 M potassium phosphate, pH 6.3, 60% (v/v) formamide in 60 min; flow-rate, 0.5 ml/min. Fraction I and fraction II represent the desired 5'-DMT-eicosadeoxynucleotide and are homogeneous, while fraction III is heterogeneous and contains only to a minor degree the desired product and mostly shorter oligodeoxynucleotides.

unlabelled oligodeoxynucleotide, which can be carried out by preparative polyacrylamide gel electrophoresis in the presence of 7 M urea (21), or more conveniently by HPLC (22). *Figure 8* shows the separation of a phosphorylated from an unphosphorylated eicosadeoxynucleotide. The radioactive phosphorylated oligodeoxynucleotide was to be used as a probe for exon III of the human *c-myc* gene. In order to achieve highest sensitivity, it was necessary to avoid competitive hybridization of the unlabelled probe. *Figure 8a* shows the separation by reversed-phase HPLC on a C_{18} column, *Figure 8b*

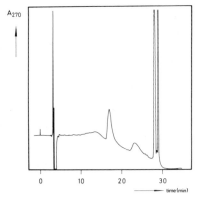

Figure 7. Separation of a phosphorylated from an unphosphorylated octadeoxynucleotide by anion-exchange HPLC on a SAX column. Sample, 350 ng each of d(GGAATTCC) and d(pGGAATTCC); column, Whatman Partisil 10 SAX (4.6 × 250 mm); apparatus, DuPont 850 liquid chromatograph with a DuPont UV spectrophotometer set at 270 nm with 0.08 AUFS; elution, linear gradient from 1 mM potassium phosphate, pH 6.3, 60% (v/v) formamide to 0.18 M potassium phosphate, pH 6.3, 60% (v/v) formamide in 30 min. flow-rate, 0.5 ml/min. d(GGAATTCC) and d(pGGAATTCC) are eluted at 28 and 29 min, respectively. Peaks at 17 and 23 min represent material carried in by the kinase reaction.

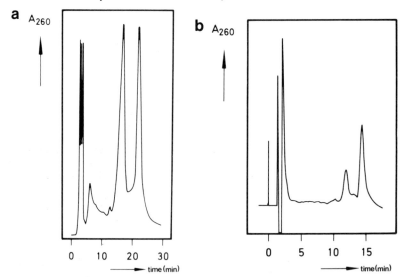

Figure 8. (a) Separation of a phosphorylated from an unphosphorylated eicosadecamer by reversed-phase HPLC on a C$_{18}$ column. Sample, ~1 μg each of d(ACAACTACGCAGCGCCTCCC) and 3 μg of d(pACAACTACGCAGCGCCTCCC); column, LiChrosorb RP-18 (4.0 × 200 mm); apparatus, Merck-Hitachi 655 A-12 with a Merck-Hitachi 655-22 spectrophotometer set at 260 nm with 0.02 AUFS; elution, linear gradient from 15 to 30% (v/v) triethylammonium acetate, pH 7.0, in 30 min; flow-rate, 1 ml/min. The phosphorylated eicosadeoxynucleotide due to its greater polarity is eluted earlier (at 17 min) from the reversed-phase column than the unphosphorylated eicosadeoxynucleotide (at 23 min). (b) Separation of a phosphorylated from an unphosphorylated eicosadecamer by anion-exchange HPLC on a DEAE column. Sample, 1 μg of d(ACAACTACGCAGCGCCTCCC) and 3 μg of d(pACAACTACGCAGCGCCTCCC); column, Nucleogen DEAE 4000 (6 × 125 mm); apparatus, DuPont 850 liquid chromatograph with a DuPont UV spectrophotometer set at 260 nm with 0.04 AUFS; elution, linear gradient from 0.3 to 0.6 M NaCl in 30 mM sodium phosphate, pH 6.0, 5 M urea in 30 min. flow-rate, 2 ml/min. The phosphorylated eicosadeoxynucleotide due to its greater charge is eluted later (at 14.5 min) from the anion-exchange column than the unphosphorylated eicosadeoxynucleotide (at 12 min).

Table 5. Separation of a radiaoctive phosphorylated oligodeoxynucleotide probe from the unphosphorylated oligodeoxynucleotide.

1. Carry out the phosphorylation reaction by incubating $10-100$ pmol oligodeoxynucleotide ($\sim 0.1-1$ μg for an eicosadeoxynucleotide) with $20-200$ pmol [γ-^{32}P]ATP with $5-50$ units of T4 kinase in 67 mM Tris$-$HCl, pH 8.0, 10 mM MgCl$_2$, 10 mM 1.4 dithioerythreitol at 37°C for 30 min in a total volume of $10-100$ μl. Leave on ice until HPLC is started.

2. Load the reaction mixture onto a LiChrosorb RP-18 column (4 × 250 mm). Perform a gradient-elution from $30-60\%$ B [solvent A: 0.1 M triethylammonium acetate, pH 7.0, 1% (v/v) acetonitrile; solvent B: 0.1 M triethylammonium acetate pH 7.0, 50% (v/v) acetonitrile] in 30 min at a flow-rate of 1 ml/min. Collect fractions of 1 ml. Measure their radioactivity (Cerenkov radiation) in a scintillation counter.

3. Combine fractions of interest. Remove the solvent by evaporation to dryness *in vacuo*. Dissolve the residue in water, evaporate to dryness *in vacuo*, repeat this procedure.

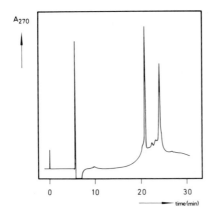

Figure 9. Separation of the reaction products, d(GGGAT) and d(pATCCC), resulting from the *Eco*RI catalysed cleavage of d(GGGATATCCC), by anion-exchange HPLC on a PEI column. Sample, 380 ng d(GGGATATCCC) digested by *Eco*RI; column, Baker PEI wide pore (4.6 × 250 mm); apparatus, Bruker LC 21 B with a Shimadzu SPD-6A spectrophotometer set at 270 nm with 0.08 AUFS; elution, linear gradient from 1 mM potassium phosphate, pH 6.3, 60% (v/v) formamide to 0.12 M potassium phosphate, pH 6.3, 60% (v/v) formamide in 30 min; flow-rate, 0.5 ml/min. Due to its greater charge d(pATCCC) is eluted later (at 24 min) from the anion-exchange column than d(GGGAT) (at 21 min).

the separation by anion-exchange HPLC on a DEAE column. The reversed-phase procedure involves only volatile eluants, such that the radioactive probe can be used directly after lyophilization (see *Table 5*). With oligodeoxynucleotides longer than approximately 20 nucleotides, anion-exchange HPLC, due to its greater resolution is superior to reversed-phase HPLC for the separation of phosphorylated from unphosphorylated oligodeoxynucleotides. In this case, however, a time-consuming de-salting step must be carried out after the HPLC runs since in anion-exchange HPLC procedures in general non-volatile buffer constituents are employed.

4.3 Analysis of restriction endonuclease catalysed cleavage of oligodeoxynucleotides

HPLC can be used to analyse reactions involving oligonucleotides, enzymatic cleavage or chemical modifications of oligonucleotides. The advantage of the HPLC analysis

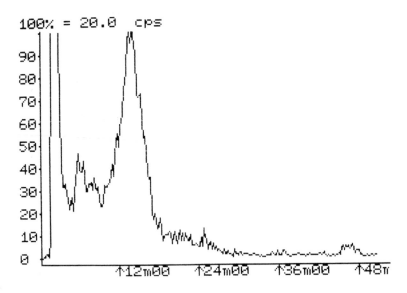

Figure 10. Anion-exchange HPLC on a DEAE column of the plasmid pETu1 into which a radioactively labelled oligodeoxynucleotide was inserted. Sample: 1.5 pmol of a synthetic 60mer and 1.5 pmol of a complementary synthetic 56mer were phosphorylated at their 5′ ends with [γ-^{32}P]ATP using T4 kinase, annealed and ligated into 0.2 pmol of the larger *Eco*RI/*Hpa*I fragment of pETu1 using ATP and T4 ligase. Half of the reaction mixture was analysed by HPLC (this figure), the other half used for transformation of competent *E. coli* cells. Column, Nucleogen DEAE-4000 (6 × 125 mm); apparatus, DuPont 850 liquid chromatograph with a Ramona-D (Isomess) flow-through radioactivity monitor; elution, linear gradient from 0.24 to 0.66 M NaCl in 10 min and then from 0.66 to 1.2 M NaCl in 120 min, in the presence of 0.03 M sodium phosphate, pH 6.0, 5 M urea. The oligodeoxynucleotide, present in large excess in the reaction mixture, is eluted at 14 min from the anion-exchange column. The plasmid carrying the insert is eluted together with the unreacted *Eco*RI/*Hpa*I fragment at 46 min. Radioactive ATP, carried into the reaction mixture with the labelled oligodeoxynucleotides, is eluted in the breakthrough.

is that there is no need for a radioactive label on the oligonucleotide and that it is very fast. *Figure 9* shows the product analysis of the *Eco*RI-catalysed digestion of d(GGGAATTCCC). Similar experiments have been carried out with modified oligodeoxynucleotides (23,24).

4.4 Analysis of the ligation of synthetic linkers into plasmid DNA

A general problem occurring in recombinant DNA work is the analysis of ligation reactions involving plasmid DNA and synthetic oligodeoxynucleotides. If the oligodeoxynucleotide is sufficiently large, the reaction can be conveniently studied by agarose gel electrophoresis. If not, one has to radioactively label the oligodeoxynucleotide, separate the educts from the products by electrophoresis or chromatography and analyse the results by autoradiography or scintillation counting. We routinely use HPLC for this purpose. This procedure is particularly elegant, when the eluate of HPLC column is analysed on line with a radioactivity monitor, as shown in *Figure 10*.

4.5 Fractionation of restriction endonuclease digests of plasmid and bacteriophage DNA

Restriction endonuclease digests of DNA are routinely analysed on agarose or

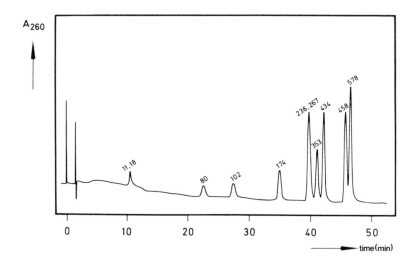

Figure 11. Semi-preparative separation of restriction fragments by anion-exchange HPLC on a DEAE column. Sample, 6.7 μg of a *Hae*III digest of pUC8; column, Nucleogen DEAE-4000 (6 × 125 mm); apparatus, DuPont 850 liquid chromatograph with a DuPont UV spectrophotometer set at 260 nm with 0.08 AUFS; elution, linear gradient from 0.24 to 0.66 M NaCl in 0.03 M sodium phosphate, pH 6.0, 5 M urea in 10 min; flow-rate, 2 ml/min. The fragments comprising 458 and 578 base pairs unlike all other ones have a high d(A+T) content which might explain their relatively long retention time.

Table 6. Purification of restriction fragments.

1.	Prepare the restriction digest according to established protocols. Stop the reaction by incubation at 95°C for 5 min or put the reaction mixture on ice until the HPLC is started. Do not use EDTA to stop the reaction since it interferes in high concentrations with the separation.
2	Load the sample (up to 10 mg, more can be accommodated but will be accompanied by a decrease in resolution) onto the Nucleogen-DEAE 4000 column (6 × 125 mm). Carry out a gradient elution from 0.25 to 1.25 M NaCl in 0.03 M sodium phosphate, pH 6.0, 5 M urea, in 120 min at a flow-rate of 2 ml/min.
3.	Combine the fractions of interest. Dialyse against water. Lyophilize the dialysate. Dissolve the residue in 0.25 ml 50 mM Tris−HCl, pH 7.5, 0.3 M sodium acetate. Add 0.75 ml ethanol. Leave on ice for 5 min. Centrifuge. Dry the pellet and dissolve it in the desired buffer.

polyacrylamide gels; there is no reasonable alternative for this procedure. For preparative purposes, however, electrophoresis is not the procedure of choice, because its capacity is low, work-up of the sample is tedious and very often contaminants are eluted from the gel which interfere with subsequent enzymatic reactions. Anion-exchange HPLC on the other hand is ideally suited for this purpose. *Figure 11* shows the semi-preparative HPLC separation of a restriction enzyme digest of a small plasmid DNA. Restriction fragments are separated according to size. A linear relationship exists between the salt activity necessary for the elution of the restriction fragment and the logarithm of the size expressed in base pairs (8). Hydrophobic interactions lead to deviations from this relationship. For example, d(A+T) rich fragments are somewhat retarded. Our procedure for the preparation of restriction fragments by HPLC is given in *Table 6*.

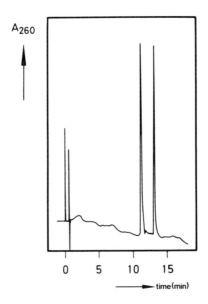

Figure 12. Analytical anion-exchange HPLC of supercoiled and linearized plasmid DNA on a DEAE column. Sample, 0.19 μg supercoiled pUC8 DNA and 0.33 μg pUC8 DNA linearized with *Eco*RI; column, Nucleogen DEAE-4000 (6 × 125 mm); apparatus, Merck-Hitachi 655 A-12 with a Merck-Hitachi 655A-22 spectrophotometer set at 260 nm with 0.08 AUFS; elution, linear gradient from 0.66 to 1.2 M NaCl in 0.03 M sodium phosphate, pH 6.0, 5 M urea in 60 min; flow-rate, 2 ml/min. Supercoiled plasmid DNA is eluted before (at 11.5 min) linear plasmid DNA (at 13.5 min). The minute peak following the peak for supercoiled plasmid DNA represents relaxed plasmid DNA.

Table 7. Preparation of supercoiled plasmid DNA.

1.	Prepare a bacterial cell lysate according to established procedures, digest the RNA with RNase, extract proteins with phenol and precipitate the DNA with polyethyleneglycol or isopropanol.
2.	Dissolve the crude DNA precipitate in 10 mM Tris−HCl, pH 8.0 at a concentration of ∼0.5 mg/ml and add 1 volume of starting buffer. Load the solution, which should contain not more than 1 mg DNA, onto a Nucleogen-DEAE 4000 column (6 × 125 mm). Carry out a gradient-elution from 0.5 to 1.25 M NaCl in 0.03 M potassium phosphate, pH 6.5, 5 M urea, in 60 min at a flow-rate of 2 ml/min. Supercoiled plasmid DNA is eluted at ∼1 M NaCl.

4.6 Separation of supercoiled, open circular and linear plasmid DNA

HPLC of nucleic acids of large molecular weight is now routinely carried out both for preparative and analytical purposes (for a review see ref. 25). In a laboratory in which HPLC is available, plasmid DNA sequencing (26) should be carried out by HPLC rather than by CsCl gradient centrifugation in the presence of ethidium bromide since the HPLC procedure is much faster and produces supercoiled plasmid DNA of high purity. *Figure 12* shows an analytical anion-exchange HPLC of supercoiled and linear plasmid DNA. The separation and work-up procedure for plasmid DNA is detailed in *Table 7*.

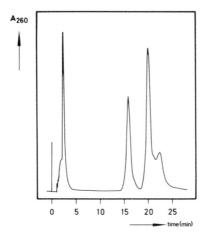

Figure 13. Isolation of single-stranded plasmid DNA by anion-exchange HPLC on a DEAE column. Sample, 500 μg of a single-stranded DNA preparation resulting from superinfection of *E.coli* LK111 cells carrying the plasmid pRIF309+ with 10 p.f.u./cell bacteriophage f1; column, Nucleogen DEAE 4000 (6 × 125 mm); apparatus, DuPont 850 liquid chromatograph with a DuPont UV spectrophotometer set at 260 nm with 2.56 AUFS; elution, linear gradient from 0.66 to 0.84 M NaCl in 0.03 M sodium phosphate, pH 6.0, 5 M urea, in 45 min; flow-rate, 2 ml/min. Single-stranded pRIF309+ DNA is eluted at 16 min. The peaks at 20.5 and at 22.5 min represent aggregates of pRIF309+ and f1 DNA.

4.7 Separation of single-stranded plasmid and bacteriophage DNAs

Plasmid DNAs carrying the bacteriophage M13 or f1 origin of replication (27) are getting increasingly important for dideoxysequencing or site directed mutagenesis, since they can be propagated as double-stranded DNA or upon infection by bacteriophage M13 or f1 as single-stranded DNA. The purity of the single-stranded DNA preparation is critical for sequencing and site directed mutagenesis work. A very fast and highly efficient method to obtain single-stranded plasmid DNA from bacteriophage infected *Escherichia coli* cultures involves anion-exchange HPLC of crude DNA preparations from the supernatant of such cultures. *Figure 13* shows a preparative separation.

In contrast to double-stranded DNAs, with single-stranded DNAs the separation is not strictly according to size, since the single-stranded DNA can engage in hydrophobic interactions with the matrix of the column packing material more readily than double-stranded DNA. Furthermore, the denaturing agent (urea or formamide) might not abolish all double-stranded stretches in the otherwise single-stranded DNA at ambient temperature; this will also effect a separation according to size.

It seems not to unrealistic to predict that with the proper choice of chromatographic conditions (concentrations of denaturing agent, temperature, pH, etc.) it will become a routine matter to achieve separation of the complementary strands of open circular or linear plasmid DNA by HPLC. Of particular interest in this respect is the MonoQ column packing material, which can tolerate alkaline pH.

4.8 Fractionation of RNase digests of tRNAs

HPLC is the method of choice for the separation of oligoribonucleotides, whether

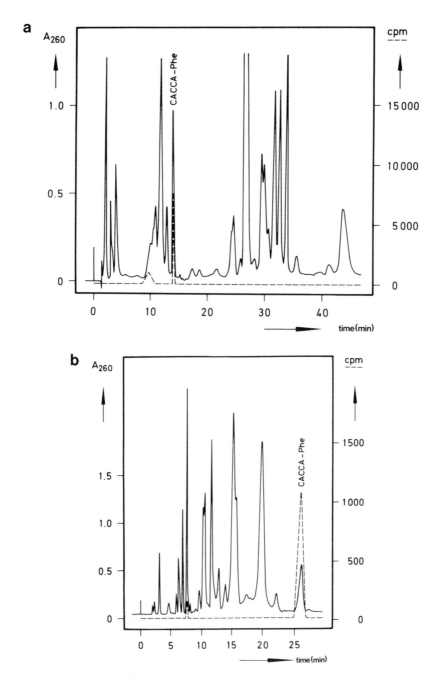

Figure 14. (a) Semi-preparative anion-exchange HPLC on a RP-NH$_2$-column of a nuclease digest of an aminoacyl-tRNA. Sample, 3 A_{260} units of a ribonuclease T1 digest of [^{14}C]Phe-tRNA ^{Phe}(yeast); column, LiChrosorb RP-NH$_2$ (4.6 × 250 nm); apparatus, DuPont 850 liquid chromatograph with a DuPont UV spectrophotometer set at 260 nm with 1.28 AUFS; elution, linear gradient from 0 to 0.1 M ammonium phosphate, pH 4.5, in 15 min, from 0.1 to 0.3 M ammonium phosphate, pH 4.5, in 10 min and from 0.3 to 1 M ammonium phosphate, pH 6.0, in 10 min; flow-rate, 1.5 ml/min. Radioactivity was monitored off-

line using a scintillation counter. The fragment of interest, 3'-*O*-Phe-r(CAACCA), due to its small charge is eluted early from the anion-exchange column. (**b**) Preparative reversed-phase HPLC on a C_{18} column of a nuclease digest of an aminoacyl-tRNA. Sample, 20 A_{260} units of a ribonuclease T1 digest of [^{14}C]Phe-tRNA^{Phe} (yeast); column, LiChrosorb RP-18 (4.6 × 250 mm); apparatus, DuPont 850 liquid chromatograph with a DuPont UV spectrophotometer set at 260 nm with 2.56 AUFS; elution, linear gradient from 0 to 10% (v/v) methanol in 5 min and from 10 to 25% (v/v) methanol in 20 min, in the presence of 0.05 M ammonium phosphate, pH 4.5; flow-rate, 4 ml/min. Radioactivity was monitored off-line by scintillation counting. 3'-*O*-Phe-r(CACCA) is eluted as the last fragment from the reversed-phase column because of the hydrophobicity of the aminoacyl residue. For this particular problem reversed-phase HPLC proved to be superior to anion-exchange HPLC (*see part a*).

obtained by enzymatic digestion of small RNA (tRNAs, small rRNAs, snRNAs, viroid RNAs, etc.), by a T4 RNA ligase-catalysed reaction, or by chemical synthesis. Both reversed-phase and anion-exchange HPLC are suitable for this purpose. *Figure 14* shows the purification of a 2'(3')-aminoacylpentaribonucleotide by HPLC on a RP-NH$_2$-column and on a C_{18} column. This procedure can be easily scaled-up for the separation of up to 100 A_{260} units of RNAs on analytical columns.

5. ACKNOWLEDGEMENTS

This chapter is dedicated to Prof. Dr Ernst Bayer (University of Tübingen) on the occasion of his 60th birthday. We gratefully acknowledge the valuable contributions of Drs J.Alves, W.Block, H.-J.Ehbrecht, R.Geiger, W.Haupt, F.-U.Gast, O.Wolf and H.Wolfes to this work. We thank Drs H.-J.Fritz, L.McLaughlin and D.Riesner for many valuable discussions. Thanks are due to Mr R.Mull for the drawings and photographs, Ms M.Krome, Mrs E.Schuchardt and Mrs L.Sell for typing the manuscript. Work in the authors' laboratories was supported by grants from the Deutsche Forschungsgemeinschaft and the Fonds der Chemischen Industrie.

6. REFERENCES

1. Gait,M.J. and Sheppard,R.C. (1977) *Nucleic Acids Res.*, **4**, 1135.
2. Newton,C.R., Greene,A.R., Heathcliffe,G.R., Atkinson,T.C., Holland,D., Markham,A.F. and Edge,M.D. (1983) *Analyt. Biochem.*, **129**, 22.
3. Haupt,W. and Pingoud,A. (1983) *J. Chromatogr.*, **260**, 419.
4. Pearson,J.D. and Regnier,F.E. (1983) *J. Chromatogr.*, **255**, 137.
5. Drager,R.R. and Regnier,F.E. (1983) *Analyt. Biochem.*, **145**, 47.
6. Colpan,M. and Riesner,D. (1984) *J. Chromatogr.*, **296**, 339.
7. Kato,Y., Sasaki,M., Hashimoto,T., Murotso,T., Fukushige,S. and Matsubara,K. (1983) *J. Chromatogr.*, **265**, 342.
8. Müller,W. (1986) *Eur. J. Biochem.*, **15**, 203.
9. Cubellis,M.V., Marino,G., Mayol,L., Piccialli,G. and Sannia,G. (1985) *J. Chromatogr.*, **329**, 406.
10. McLaughlin,L.W. and Krusche,J.U. (1982) In *Chemical and Enzymatic Synthesis of Gene Fragments: A Laboratory Manual.* Gassen,H.G. and Lang,A. (eds), Verlag Chemie, Weinheim, p. 177.
11. Fritz,H.-J., Belagaje,R., Brown,L.E., Fritz,R.H., Jones,R.A., Lees,R.G. and Khorana,H.G. (1978) *Biochemistry,* **17**, 1257.
12. Fritz,H.-J., Eick,D. and Werr,W. (1982) In *Chemical and Enzymatic Synthesis of Gene Fragments: A Laboratory Manual.* Gassen,H.G. and Lang,A. (eds), Verlag Chemie, Weinheim, p. 199.
13. Kelmers,A.D., Novelli,D.G. and Stulberg,M.P. (1965) *J. Biol. Chem.*, **240**, 3979.
14. Wells,R.D. (1984) *J. Chromatogr.*, **336**, 3.
15. Bischoff,R., Graeser,E. and McLaughlin,L.W. (1983) *J. Chromatogr.*, **257**, 305.
16. Floyd,T.R., Cicero,S.E., Fazio,S.D., Raglione,T.V., Hsu,S.-H., Winkle,S.A. and Hartwick,R.A. (1986) *Anal. Biochem.*, **154**, 570.
17. Molko,D., Derbyshire,R., Guy,A., Roget,A., Teoule,R. and Boucherle,A. (1981) *J. Chromatogr.*, **206**, 493.

18. Kruppa,J., Graeve,L., Bauche,A. and Földi,P. (1984) *LC Magazine,* **2**, 848.
19. McLaughlin,L.W. and Piel,N. (1984) In *Oligonucleotide Synthesis: A Practical Approach.* Gait,M.J. (ed.), IRL Press, Oxford, p. 207.
20. Becker,C.R., Efcavitch,J.W., Heiner,R.C. and Kaiser,N.F. (1985) *J. Chromatogr.,* **326**, 293.
21. Thein,S.L. and Wallace,R.B. (1986) In *Human Genetic Diseases: A Practical Approach.* Davies,K.E. (ed.), IRL Press, Oxford, p. 33.
22. Delort,A.M., Derbyshire,R., Duplaa,A.M., Guy,A., Molko,D. and Teoule,R. (1984) *J. Chromatogr.,* **283**, 462.
23. Connolly,B.A., Potter,B.V.L., Eckstein,F., Pingoud,A. and Grotjahn,L. (1984) *Biochemistry,* **23**, 3443.
24. Seela,F. and Driller,H. (1986) *Nucleic Acids Res.,* **14**, 2319.
25. Colpan,M. and Riesner,D. (1988) *Comprehensive Biochemistry,* in press.
26. Chen,E.Y. and Seeburg,P.H. (1985) *DNA,* **4**, 165.
27. Dente,L., Ceserini,G. and Cortese,R. (1983) *Nucleic Acids Res.,* **11**, 1645.

CHAPTER 8

Affinity chromatography of complex carbohydrates using lectins

JOHN T.GALLAGHER

1. INTRODUCTION

In recent years affinity chromatography has made a dramatic impact upon preparative and analytical biochemical separation procedures for complex biological mixtures. Although it is a form of adsorption chromatography, the technique is unique because it exploits the naturally-occurring binding specificities that can be found in living systems. Good examples are the interactions between antibodies and antigens, enzymes and substrates or receptors and ligands. Experimentally, affinity chromatography requires one component of a bi-molecular interaction to be immobilized on an insoluble support, such as an agarose gel, and the derivatized support is then packed into a small column. When a solution containing a complex mixture of substances is passed through the column, only molecules with the appropriate binding specificity will be retained, and these may then be eluted using suitable dissociating solutions.

2. LECTINS AS AFFINITY CHROMATOGRAPHIC SUPPORT MATERIALS

2.1 General features

Lectins are proteins, widely distributed in the plant and animal kingdoms, that have the distinctive property of binding to carbohydrates of specific structure and configuration (1,2,3,4,4a). In the majority of cases lectins are composed of identical subunits, each subunit containing one sugar-binding site. There are some exceptions, including the widely used lectin wheat germ agglutinin (WGA), a dimeric protein in which each subunit contains two binding sites (5). The multivalency of lectins is responsible for their characteristic ability to agglutinate cells or glycoconjugates and very high affinity interactions may occur with complex glycopeptides that contain two or more lectin-binding domains.

Lectins are ideally suited to affinity chromatography. They can be coupled efficiently to an affinity matrix with retention of their biological activity. If properly maintained, they can be used repeatedly and stored for prolonged periods. Lectins can be selected that display considerable differences in their carbohydrate-binding properties and careful use of appropriate combinations of lectin affinity columns can provide rapid and extensive information on oligosaccharide structures in heterogeneous samples (2,6). At the other extreme, lectins can identify minimal differences between carbohydrates. For example, they can be used to differentiate epimers of pyranose sugars (e.g. mannose and galactose) or to determine the anomeric configuration of a specific monosaccharide (see *Tables 1* and *2*).

Table 1. Lectin-binding to N-linked oligosaccharides.

Oligosaccharides	Lectin				
	Con A	LCA	L-PHA	E-PHA	WGA

(a) High mannose

```
M —α1,2— M
            \ α1,6
M —α1,2— M —α1,2— M
                      \ α1,6
                        M —β1,4— O —β1,4— O —— Asn
                      / α1,3
M —α1,2— M —α1,2— M
```

| | + | − | − | − | − |

(b) Complex, bi-antennate

```
          α2,3(6)      β1,4    β1,2
NeuAc ——————— Gal ——— O ——— M
                                 \ α1,6
                                   M —β1,4— O —β1,4— O —— Asn
                                 / α1,3
          α2,3(6)      β1,4    β1,2
NeuAc ——————— Gal ——— O ——— M
```

| | + | − | − | − | − |

(c) Complex, bi-antennate-fucosylated

```
NeuAc — Gal — O — M
                    \
                      M — O — O — Asn
                    /       | α1,6
NeuAc — Gal — O — M         F
```

| | + | + | − | − | − |

(d) Complex, tri-antennate-fucosylated

```
NeuAc — Gal — O
                \ β1,6
NeuAc — Gal — O — M
                    \
                      M — O — O — Asn
                    /       |
NeuAc — Gal — O — M         F
```

| | − | + | + | − | − |

(e) Complex, tetra-antennate

```
NeuAc — Gal — O
                \ β1,6
NeuAc — Gal — O — M
                    \
                      M — O — O — Asn
                    /
NeuAc — Gal — O — M
                / β1,4
NeuAc — Gal — O
```

| | − | − | + | − | − |

(f) Complex, tri-antennate, bisected

```
                     O
Gal — O — M          | β1,4
            \        
              M — O — O — Asn
            /
Gal — O — M
        / β1,4
Gal — O
```

| | − | − | − | + | + |

(g) Hybrid, bisected

(h) Polylactosamine

(i) O-linked, mucin type;

O, N−acetylglucosamine (GlcNac); F, fucose; M, mannose; Gal, galactose; NeuAc, N-acetylneuraminic acid (sialic acid).

Table 2. Examples of monosaccharide inhibitors of lectins.

Lectin	Inhibitory sugars
Con A	α-Man
Lentil lectin (LCA)	α-Man
Pea lectin	α-Man
Griffonia simplicifolia 2	GlcNAc
Helix pomatia (snail)	α-D-GalNAc
Griffonia simplicifolia 1A	α-D-GalNAc
Vicia vellosa	α-D-GalNAc
Ulex europeus (Gorse)	α-L-Fuc
Lotus tetragonolobus	α-L-Fuc
Griffonia simplicifolia-1B	α-D-Gal
Ricinus communis agglutinin-I (RCA-I)	Gal
Ricinus communis agglutinin-II (RCA-II)	Gal
Arachis hypogea (peanut PNA)	Gal or lactose
Triticum vulgaris	Chitobiose (GlcNAc)$_2$
Limax flavus	NeuAc

Sugars are normally more inhibitory when used as their methyl glycosides. It is clearly advantageous to use a methylglycoside with the appropriate anomeric configuration for those lectins which show a strong anomeric preference. For example, α-methylmannoside is a more potent inhibitor of Con A and LCA than free mannose and it is therefore the most effective derivative for eluting oligosaccharides from Con A and LCA affinity columns. With the exception of WGA, all the lectins shown are exolectins (class I) which are readily eluted from affinity columns by monosaccharides.

2.2 Carbohydrate-binding properties of lectins

It is important to appreciate the range and complexity of the carbohydrate-binding properties of lectins if full advantage is to be taken of the opportunity offered by lectin affinity chromatography. For classification purposes, the author has proposed a scheme

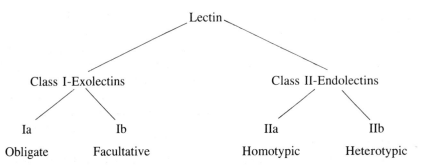

Figure 1. Lectins. Classification and nomenclature. Details are given in the text. Briefly, exolectins (class I) show a primary specificity for monosaccharides and exolectin binding to complex oligosaccharides can be inhibited by single sugars. Exolectins bind to end-chain sugars though some, the facultative exolectins, will also recognize internal sugars. The endolectins (class II) interact only with specific sugar sequences which may be composed of identical sugars (homotypic sequences) or of two or more different monosaccharide units (heterotypic sequences). The scheme is described in detail in ref. 2. Of the lectins discussed in detail in the text and in *Table 1*, LCA and Con A are class Ib or facultative exolectins that recognize internal and external mannose residues, RCA is a class Ia lectin, E- and L-PHA are class IIb−heterotypic endolectins and WGA is a class IIa lectin (homotypic endolectin) though it displays strong heterotypic features through an ability to recognize Gal−GlcNAc repeat sequences (*Table 1* and Section 3.8).

recently that should be of value in helping the reader to understand the different mechanisms of carbohydrate recognition by lectins. In this scheme (2), two groups of lectins are recognized: the class I or exolectins and the class II or endolectins (*Figure 1*). The class I lectins show a primary specificity for a single sugar in a complex oligosaccharide and they will always recognize that sugar when it is found at the non-reducing end of a carbohydrate chain, hence the name exolectin. Some exolectins have a mandatory requirement for end-chain sugars and are called the 'obligate exolectins' (class Ia) whilst others recognize both peripheral and internal sugars and are described as 'facultative exolectins' (class Ib). Exolectins are most commonly used in affinity chromatography because bound glycopeptides can be displaced readily by elution with monosaccharides. However, binding and specific desorption may be indicative of considerably more structural information than the presence of a particular sugar residue. This is because the affinity of exolectins for complementary sugars may be enhanced by adjacent sugar sequences (synergistic sequences). Some of these sequences are described in *Table 1*, and a more comprehensive listing can be found in ref. 2. Desorption conditions for an affinity matrix can be chosen to distinguish between low and high affinity oligosaccharides.

The endolectins are an interesting group of proteins which display a more complex mode of carbohydrate binding than the exolectins. Endolectins bind to specific carbo-hydrate sequences but no individual sugar in these sequences plays a predominant role in the binding process. It follows that an interaction with one or more internal sugars is essential for carbohydrate recognition (endo-recognition). The homotypic endolectins (class IIa) recognize sequences of identical sugar units, the heterotypic endolectins (class IIb) bind most strongly to sequences composed of two or more different mono-saccharides. There are some important practical implications when working with these lectins. Extra care must be taken in coupling endolectins to affinity matrices since it is often impossible to obtain sufficient quantities of complementary oligosaccharides to protect the sugar-binding site during coupling (see Section 3.5). Elution of bound oligosaccharides can also cause a problem, again because competing oligosaccharides are not available. In such cases, pH gradients may be used for carbohydrate desorption. Because this method of elution does not establish binding specificity, it is important to ensure that mild procedures giving a low degree of coupling are used in the preparation of endolectin affinity matrices (Section 3.5). The great advantage of many endolectins is that they provide very specific information on sequence and branching of oligo-saccharides (*Table 1*). This is a reflection of their stringent structural requirements for saccharide recognition.

Table 1 shows a range of N-linked oligosaccharides and their reactivities with certain lectins. The trimannan−chitobiosyl core is common to all structures but for simplicity the sugar linkages in this core region are shown only for the high mannose and complex bi-antennate components (a and b). The trisaccharide NeuAc−Gal−GlcNAc forming the antennae in the complex N-glycans has the sugar linkages illustrated in the bi-antennate component (b). The polylactosamine sequences (h) can occur in any complex N-glycan, but most commonly in the tri- and tetra-antennate derivatives. These sequences are attached directly to the core mannose and they may be terminated by sialic acid. NeuAc−(Gal−GlcNAc)$_n$−Man−polylactosamine sequence may be linear (as shown) or branched:

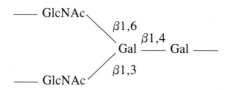

Branched and linear polylactosamines will bind to WGA (10), with binding strength correlating directly with molecular size. Polylactosamines reduce or abolish saccharide interactions with LCA, Con A, E-PHA and L-PHA (phytohaemagglutinin-erythrocytes and -lymphocytes). The complex N-glycans may lack one or more of their terminal sialic acids and the exposed galactose units will interact strongly with RCA-I or RCA-II (Sections 3.8 and 3.9). The RCA lectins, especially RCA-II, will interact with some of the sialylated structures shown in *Table 1* but with decreased affinity (10a). On some occasions, the antennae may contain terminal α-Gal or α-GalNAc units. *Griffonia simplicifolia* isolectin 1-B can be used to detect α-Gal terminals and the terminal α-GalNAc units can be identified with several exolectins including *Griffonia simplicifolia* isolectin 1-A and the *Vicia vellosa* lectin (*Table 2*). It is important to note the differences in lectin binding between LCA and Con A. Both lectins are inhibited by α-MM (*Table 2* and *Figures 2* and *3*) but LCA requires a core fucose residue (structures c and d) and will not bind to the high mannose or hybrid glycans which interact with Con A with high affinity. However LCA binds to tri-antennate glycans in which the α1,6-linked core mannose is substituted at C-2 and C-6 with GlcNAc (d). The differences in binding properties between these two lectins can be usefully exploited in serial affinity chromatography (*Figure 2*) LCA and Con A both show higher affinity for N-glycans with short antennae with GlcNAc or galactose terminals (see refs 12,25 for further

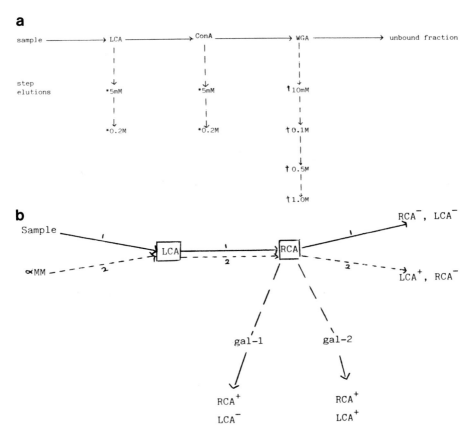

Figure 2. (a) Serial lectin affinity chromatography. Direct collection of desorbed material. Samples are applied to the first column in the series in equilibration buffer and eluted through the three affinity columns (Con A, LCA, WGA) connected in series. The columns are then disconnected and step-eluted with appropriate sugars as indicated.*, concentration of α-methylmannoside; †, concentration of GlcNAc. **(b)** Serial lectin affinity chromatography. Direct application of desorbed material to a second affinity column. Samples are applied in equilibration buffer and non-binding glycopeptides (RCA⁻, LCA⁻) are collected. The columns are disconnected and the RCA column eluted with 0.1 M galactose (gal-1) to yield RCA-binding glycopeptides (RCA⁺) that have no affinity for LCA (LCA⁻). The columns are then re-connected and eluted with α-MM (5 mM, - - - -), this desorbs LCA-binding glycopeptides (LCA⁺) which then pass onto the RCA column. LCA⁺ glycopeptides that lack terminal gal residues are not retarded and emerge as LCA⁺, RCA⁻. The columns are disconnected again and the RCA column eluted for a second time with 0.1 M galactose (gal-2) to yield RCA⁺ LCA⁺. The columns may be reconnected and the process repeated with a higher concentration of α-MM.

details). The ability of the iso-endolectins, L-PHA and E-PHA to recognize distinct branching patterns on the mannan core is striking. L-PHA has a mandatory requirement for the core α1,6-linked mannose to be substituted at C-2 and C-6 with GlcNAc residues. It shows higher affinity for tetra-antennate compared to tri-antennate oligosaccharides (15,16). In contrast, E-PHA requires a bisected glycan in which a β1,4 GlcNAc is attached to the β-mannose of the glycan core. The bisected tri-antennate structure shown in *Table 1* is the most complementary oligosaccharide for E-PHA that has so far been identified (17). Note that L-PHA will not bind to these bisected glycans and E-PHA

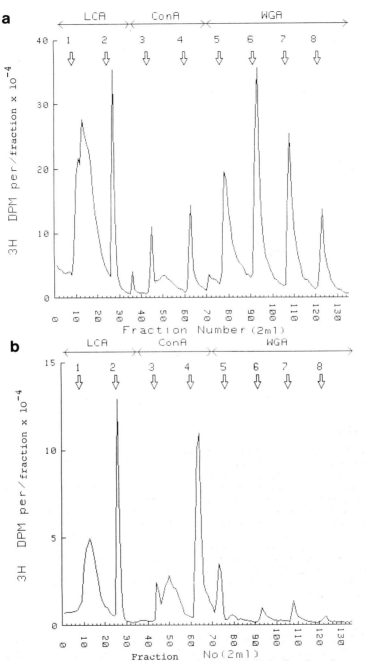

Figure 3. Serial lectin affinity chromatography of metabolically-labelled cell surface glycopeptides from a murine haemopoietic cell line. (**a**) [³H]Glucosamine; (**b**) [³H]mannose. The cells were incubated for 48 h with either [³H]glucosamine or [³H]mannose. Washed cells were treated with trypsin to release surface glycopeptides which were then treated with pronase. The pronase–glycopeptides were then analysed by serial lectin affinity chromatography as described in *Figure 2a*. Note the contrast in radiolabelling of WGA-binding glycopeptides between the two isotopic precursors and the high concentration of [³H]mannose in the high affinity Con A fraction. Arrows indicate the step elutions with competing sugars—these were 5 mM and 0.2 α-MM for LCA and Con A and 10 mM, 0.1 M, 0.5 M and 1.0 M GlcNAc for WGA (10,26).

215

will not bind to structures that contain C-2 and C-6 substituted mannose units that are vital for the interaction with L-PHA. E- and L-PHA will bind to sialylated structures especially if the sialic acid is linked $\alpha,2,3$, but binding is stronger with unsialylated or partially sialylated components.

3. LECTIN AFFINITY CHROMATOGRAPHY

3.1 **Sample preparation**

There are two basic types of carbohydrate−protein linkages in glycoproteins, GlcNAc−Asn or 'N'-glycosidic linkages and the GalNAc−Ser (or Thr) or O-glycosidic linkages. Xylose−Ser, O-glycosidic bonds occur in proteoglycans.

The origin of glycopeptides or oligosaccharides can vary considerably from highly purified glycoproteins to trypsin extracts of cell surfaces. The latter types of sample will be very heterogeneous. Oligosaccharides from purified glycoproteins will be less variable but could still show microheterogeneity with respect to peripheral sugar units.

In the majority of cases glycopeptides containing single N-linked oligosaccharide chains can be prepared by digesting glycoproteins with a broad spectrum protease such as pronase.

Such samples may also be treated with the enzyme PGNaseF (sometimes called N-glycanase) which will degrade the N-glycosylamine linkage between the saccharide chain and asparagine of the protein core (7). Alternatively this linkage may be cleaved by hydrazinolysis (8).

Oligosaccharides linked to serine or threonine in the protein core (O-linked oligo-saccharides) often occur in clusters along a central polypeptide (10). Proteases cannot degrade the polypeptide between the attachment points of the individual sugar chains. Therefore proteolysis of glycoproteins containing O-linked oligosaccharides often produces large glycopeptides with multiple saccharide chains. These can be released by base−borohydride elimination. It may be useful to compare the lectin-binding properties of oligosaccharide clusters with the corresponding free chains. Methods for oligosaccharide and glycopeptide preparations are given in *Table 3*.

Table 3. Preparation of glycopeptides and oligosaccharides.

---

1. *Pronase digestion*

(i)   Dissolve samples in 0.1 M Tris−acetate buffer, pH 7.8 containing 0.15 M NaCl and 1 mM calcium acetate to give a sample concentration of 5% w/v.

(ii)   Add pronase (0.25% w/v) and incubate the solution for 24 h at 37°C in the presence of a few drops of toluene.

(iii) Adjust the pH, if necessary, to 7.8, add additional pronase (0.1% w/v) and continue the incubation for a further 24 h at 37°C.

(iv)   Cool the solution and add ice-cold 100% w/v trichloroacetic acid (TCA) to give a final concentration of 10% w/v TCA.

(v)   Stand for 1 h at 4°C, then remove the protein precipitate by centrifugation at 2000 r.p.m. for 10 min.

(vi)   Dialyse the supernatant exhaustively against distilled water at 4°C using low mol. wt cut-off dialysis tubing and concentrate the glycopeptides by rotary evaporation and freeze-drying.

If only very small amounts of material are available (this is often the case with metabolically-labelled glyco-peptides from cell cultures) and the sample weight cannot be determined accurately it is usually adequate to incubate the material overnight with 2.5 mg/ml pronase. Thereafter the procedure could follow that described above or, since only very small quantitites of protein are present in the sample, the pronase could be inactivated by heating at 100°C for 10 min and samples applied directly to lectin affinity columns.

2. *Peptide:N-glycosidase F (PNGase F)*

This enzyme will hydrolyse all types of oligosaccharide that are linked to protein through a GlnNAc−Asn, N-glycosidic bond. Susceptibility to the enzyme is greatly enhanced if glycoproteins are first denatured with sodium dodecyl sulphate (SDS).

The following method is taken from Tarentino *et al.* (7).

(i)   Make up glycoprotein stocks (5 mg/ml) in 1% w/v SDS and boil for 3 min.

(ii)   Add 50 μg of denatured glycoprotein to 100 μl of 0.25 M sodium phosphate, pH 8.6 containing 0.6% NP-40 and 6 milliunits[a] of enzyme.

(iii) Incubate samples for 18 h at 37°C. The enzyme can then be heat inactivated and the samples applied directly to lectin affinity columns.

The susceptibility of glycoproteins to the enzyme may vary so the above conditions may have to be modified somewhat to achieve complete de-glycosylation.

It should be noted that this method yields free oligosaccharides, pronase yields glycopeptides. Pronase-derived glycopeptides can be treated with PNGase F to liberate oligosaccharides. In this way, the influence of the peptide moiety in carbohydrate-lectin interactions can be studied.

PGNase F is often marketed under the name endo F (endo-β-*N*-acetyl glucosaminidase F). Endo F and PGNase F are separate enzymes. Endo F cleaves the hexosaminidic linkage in the chitobiosyl core of N-linked glycans. Though its activity is largely suppressed at the high-pH conditions used for PGNase F, some hydrolysis of the terminal core GlcNAc units will occur.

3. *Hydrazinolysis*

This method can be used to de-glycosylate glycoproteins but the reagents require very careful handling. The procedure is described in ref. 8.

4. *Alkaline borohydride elimination*

Some carbohydrate chains are linked to protein by O-glycosidic linkages between serine or threonine in the protein core and GalNAc in the oligosaccharide. These linkages are common in mucins and may be found in membrane glycoproteins. They often occur in clusters along the polypeptide and pronase treatment often fails to hydrolyse all of the peptide links in these heavily glycosylated domains. These oligosaccharides can be liberated by a base−borohydride elimination reaction as follows.

(i)   Take up samples (5−10 mg/ml) in 0.1 M NaOH containing 1 M $NaBH_4$ for 48 h at 37°C under an N_2 atmosphere.

(ii)   Neutralize the solution by addition of 1 M acetic acid, using phenol red as indicator.

(iii) Equilibrate the oligosaccharide with the affinity chromatography buffer by gel filtration on Bio Gel P.2 or Sephadex G-15.

---

[a]One milliunit is defined as 1 nmol of [^{3}H]dansylated pentaglycopeptide−[^{3}H-Dns-Asn(GlcNAc)$_2$(Man)$_3$] hydrolysed per min.

The method of sample preparation has a profound influence on the binding activities of the two isolectins, phytohaemagglutinin-erythrocytes (E-PHA) and phyto-haemagglutinin-lymphocytes (L-PHA). The distinctive binding properties of these lectins, that are described in *Table 1*, apply to glycopeptides that are prepared from glycoproteins by treatment with pronase. However, in a study of complex oligo-saccharides released from glycoproteins by PNGase (N-glycanase), that lack the peptide component of glycopeptides (10a), the differences in binding activity between these two lectins are abolished. Both lectins bind to bisected oligosaccharides, which are only recognized by E-PHA when presented in the form of a glycopeptide, and to tetra-antennate oligosaccharides, that as glycopeptides bind only to L-PHA (see *Table 1* for details).

217

Table 4. Chemical or enzymatic techniques for radiolabelling glycopeptides.

1. *Sialic acid*

The hydroxyl groups in the acyclic carbons (C7−C9) of sialic acid may be specifically oxidized by dilute sodium periodate (NaIO$_4$) to yield a seven-carbon sugar with a C7 aldehyde which can be reduced with NaB^{3}H$_4$.

(i)   Dissolve sufficient glycopeptides containing 1 mol of sialic acid in 1 ml of 0.1 M sodium acetate, 0.15 M NaCl pH 5.6 and cool to 0°C.
(ii)   Add an equal volume of ice-cold 10 mM NaIO$_4$ and keep the mixture at 0°C in the dark for 10 min.
(iii)   Stop the reaction by adding 50 μl of glycerol and dialyse the solution at 4°C against 50 mM sodium phosphate, 0.15 M NaCl, pH 7.5.
(iv)   Add 15 μl of NaB^{3}H$_4$ (200−300 mCi/mmol) in 10 mM NaOH and, after 30 min at room temperature, add a further 50 μl of unlabelled 0.1 M NaBH$_4$ in 10 mM NaOH to ensure complete reduction.

Excess NaBH$_4$ can be removed from the [^{3}H]glycopeptides by dialysis or gel filtration.
Note: Experiments with NaB^{3}H$_4$ should be carried out in a fume cupboard.

2. *Galactose*

Galactose residues (and GalNAc units) can be labelled using galactose oxidase/NaB^{3}H$_4$. The enzyme mainly attacks end-chain galactose. If labelling is poor, it can be increased by prior treatment of the glycopeptides with neuraminidase.

(i)   Incubate glycopeptides (up to 0.5 mg/ml), galactose oxidase (12 units/ml) and horseradish peroxidase (18 units/ml) for 48 h at 25°C in 20 mM sodium phosphate, 0.45 M sodium acetate, 0.15 M NaCl, pH 7.0.
(ii)   Add a drop of toluene to inhibit microbial growth.
(iii)   Stop the reaction by careful addition of 100% w/v ice cold TCA to a final concentration of 10% w/v TCA.
(iv)   After standing at 0°C for 1 h centrifuge the sample at 800 *g* for 5 min, dialyse the supernatant against distilled water and concentrate to about 1 ml by rotary evaporation.
(v)   Dialyse the sample against 50 mM sodium phosphate, 0.15 M NaCl, pH 7.5 and reduce with NaB^{3}H$_4$ in 10 mM NaOH as described for sialic acid labelling.

3. [^{14}C]*Acetylation of glycopeptides*

(i)   Incubate 0.5 μmol glycopeptides with 1.5 μmol of [1-^{14}C]acetic acid anhydride (5−10 mCi/mmol) in 50 μl of 0.5% w/v NaHCO$_3$ at room temperature for 1 h.
(ii)   Add 5 μmol of unlabelled acetic acid anhydride and continue the incubation for another hour.
(iii)   With care, repeatedly flash evaporate the solution to dryness from water and finally resuspend it in 1 ml of water.

Gel chromatography on Biol Gel P4 separates the [^{14}C]polypeptides from low mol. wt [^{14}C]acetate that was not removed by evaporation (25).
Note: The radiolabelled acetic acid anhydride should be handled in a fume cupboard.
Methods are also available for ^{3}H-labelling the reducing end of free oligosaccharides (20).

3.2 **Radiolabelled samples**

Detection of glycopeptides is simplified if the samples can be labelled with radioisotopes. It is possible to metabolically label complex carbohydrates with isotopic precursors and these may provide some useful structural information. A glucosamine precursor will specifically label N-acetylglucosamine (GlcNAc), N-acetylgalactosamine (GalNAc) and sialic acid whereas N-acetyl-mannosamine will only label sialic acid (19). Radio-active mannose is incorporated as mannose and fucose. A radiolabelling ratio of 3:1, mannose:fucose is found in N-linked glycans that bind to LCA (12). Mannose is a very useful selective precursor because in mammalian cells it is found only in N-linked sequences: the high-mannose type structures (*Table 1*) label particularly well when cells are incubated with radioactive mannose (*Figure 3b*).

Chemical radiolabelling is also a very effective means of 'tagging' glycopeptides. Experimental procedures are given in *Table 4*. ^{3}H-Labels can be introduced into sialic acid and galactose at the non-reducing end of an oligosaccharide or into GlcNAc and GalNAc at the reducing terminus. Amino groups in the peptide portion of a glycopeptide can be labelled with [^{14}C]- or [^{3}H]acetic anhydride.

In the author's laboratory, the elution of radiolabelled samples is measured by liquid scintillation counting. Aliquots of collected fractions are mixed wtih OptiPhase 'Safe' (LKB) liquid scintillation fluid using a maximum of 1 vol. of sample to 10 vols of scintillant.

If samples cannot be radiolabelled then chemical analysis can be carried out for neutral sugars, amino sugars, sialic acid or assays such as the phenol−sulphuric acid method can be used for general carbohydrate analysis. These methods have recently been reviewed thoroughly by Beeley (20).

3.3 Choice of affinity matrix

An affinity matrix should have the following properties.

(i)   Stability.
(ii)   Good flow properties.
(iii)   High protein-binding capacity.
(iv)   Retain carbohydrate-binding properties of coupled lectins.
(v)   Chemically inert after coupling.

Agarose gels (sold as Sepharose gels by Pharmacia) meet these requirements and are suitable for most applications. The more commonly used lectins are commercially available as agarose or Sepharose conjugated derivatives. These are generally reliable but thorough washing in 0.5 M NaCl is recommended to remove unbound lectin. However, it is often desirable to prepare your own lectin affinity gels using matrices that have been pre-activated for direct coupling of proteins. This could be necessary for a number of reasons.

(i)   Cost-considerations when large quantities of affinity gel are required.
(ii)   Concentration of coupled lectin must be varied.
(iii)   Degree of coupling (number of attachment points between the lectin and the affinity matrix) must be low.

Some matrices (e.g. CH−Sepharose) are supplied with a spacer arm to which the lectin may be attached. These can be useful as they prevent tight adsorption to the matrix itself and the lectin may be more accessible to interacting glycopeptides. Some examples of procedures for matrix coupling are given below. It cannot be emphasized strongly enough that the key to good affinity chromatography lies in the preparation of a matrix derivatized with the appropriate concentration and degree of coupling of biologically active lectin.

3.4 **Preparation of lectin affinity gels**

Examples are given for the coupling of concanavalin A (Con A). The methods are generally applicable to other lectins. During the coupling process appropriate saccharides must be added to inhibit formation of linkage groups in the immediate vicinity of the sugar binding site.

3.4.1 *Coupling to CNBr – Sepharose*

A procedure that yields 6 – 8 mg of Con A per ml bed volume of affinity gel is described.

(i)   Weigh out 2.5 g of freeze dried CNBr-activated Sepharose 4B (1 g of dry gel gives 3 – 3.5 ml of swollen gel). Sprinkle onto 50 ml of 1 mM HCl and allow the gel to hydrate and settle for 5 min. Pour the gel onto a small sintered glass funnel (G.3) attached to a vacuum flask and draw off the excess fluid. Make six successive additions of 50 ml of 1 mM HCl drawing off surplus fluid each time.

(ii)   Prepare a solution of Con A in a stoppered glass tube, by adding 100 mg of lectin to 15 ml of coupling buffer – 0.1 M sodium bicarbonate buffer pH 7.8 containing 0.5 M NaCl and 0.05 M methyl mannoside. Measure the absorbance at 280 nm. This value will be used to determine the efficiency of the coupling of lectin to the Sepharose gel. A 1:3 dilution of the initial lectin solution has an absorbance at 280 nm of $\simeq 2.17$ (1 cm path length).

(iii) Immediately after washing transfer the swollen gel to the above Con A solution and mix gently, end-over-end for 2 h at 4°C. (Do not stir as this may cause some breakdown of the gel beads.)

(iv)   Filter the gel slurry on the sintered glass funnel and measure the absorbance at 280 nm of the filtrate to ensure that efficient coupling has occurred. A reduction of the absorbance at 280 nm by at least 50% indicates that at least 50% of the lectin has been coupled. Block the excess of active groups in the gel by mixing end-over-end for 2 h at room temperature in coupling buffer containing 0.2 M glycine.

(v)   Excess blocking agent and lectin are then removed by alternate washing on a glass filter with 50 ml of 0.1 M sodium acetate buffer pH 4.0 and 50 ml of coupling buffer (without the α-methyl mannoside). Each of these buffers contain 1 mM $CaCl_2$ and 1 mM $MnCl_2$ to ensure saturation of the metal ion binding sites in the lectin. Repeat the washes five times to ensure complete removal of unbound lectin.

(vi)   The affinity adsorbent may be stored at 4°C in 10 mM Tris – acetate buffer, pH 7.4 containing 0.05% w/v sodium azide. The gel may be damaged by freezing.

3.4.2 *Coupling to CH – Sepharose 4B (method taken from ref. 21)*

Single-step, carbodiimide-mediated, coupling to CH – Sepharose can be carried out at pH 4.5 in distilled water. Con A is a tetramer above pH 6.0 but dissociates to form dimers at lower pH values. Coupling of dimeric Con A to spacer arms as found in CH – Sepharose may have some advantages. For example it may be easier to dissociate glycoproteins that bind with high affinity to Con A from the dimer form of the lectin.

It is also possible that steric hindrance could impair the binding of large molecules to the native Con A tetramer linked directly to an affinity matrix (21). The method described gives near-quantitative coupling of Con A to the gel. The degree of coupling is likely to be lower than with CNBr−Sephadex.

(i)   Add 2.5 g of CH−Sepharose 4B to 200 ml of 0.5 M NaCl and allow to swell overnight at 4°C. Wash the gel 3−4 times with 100 ml of 0.5 M NaCl on a sintered glass funnel and finally with two washes of 200 ml of distilled water.

(ii)   Dissolve 100 mg of Con A in 10 ml distilled water and carefully adjust the pH to 4.5 by addition of 0.1 M NaOH or 0.1 M HCl.

(iii)   Mix the washed gel with the lectin solution and adjust the volume with distilled water at 27 ml.

(iv)   Add 3 ml of carbodiimide (1 M in distilled water) dropwise and mix overnight end-over-end at room temperature, carefully maintaining the pH at 4.5 for the first 2 h and making occasional pH adjustments thereafter.

(v)   Filter the affinity gel on a sintered glass disc and wash with 200 ml of water, 200 ml of 1 M NaCl in 50 mM sodium phosphate buffer, pH 8.0, followed by 200 ml of 1 M NaCl in 0.1 M sodium acetate pH 4.5 containing 1 mM $CaCl_2$ and 1 mM $MnCl_2$. Finally, wash thoroughly with distilled water. Store the gel in 0.1 M sodium acetate buffer, pH 4.5, in 0.05% w/v sodium azide at 4°C.

3.4.3 *Coupling to polyacrylic hydrazide Sepharose (adapted from ref. 22)*

The preparation of this gel does not introduce charged groups into the matrix.

(i)   Suspend the polyacrylic hydrazide Sepharose gel in distilled water (10 ml packed gel/15 ml water). Stir gently and add 5 ml of 50% (w/v) glutaraldehyde. Stand for 4 h at 4°C, then wash extensively with distilled water until the odour of glutaraldehyde has completely disappeared.

(ii)   Prepare a solution of Con A, 5 mg/ml in 0.1 M sodium acetate buffer, pH 6.8, containing 1 M NaCl and 50 mM α-methyl mannoside and measure the absorbance at 280 nm.

(iii)   Add 10 ml of the lectin solution to the gel (10 ml packed bed volume) and add an extra 10 ml of the above buffer.

(iv)   Mix end-over-end for 18 h at 4°C.

(v)   Filter the gel on a sintered glass funnel and determine the protein concentration in the filtrate from the absorbance at 280 nm. Coupling efficiency should be around 80% as measured optically.

(vi)   Suspend the affinity gel in 3 vols of 5 mM sodium phosphate, 0.1 M NaCl, pH 7.2 and add enough solid $NaBH_4$ to give a final concentration of 0.5 mg per ml. Mix and stand at 4°C for 3 h.

(vii)   Wash thoroughly in 5 mM sodium phosphate buffer containing 0.1 M NaCl and store at 4°C in this buffer in the presence of sodium azide (0.05% w/v).

3.5 High-pressure affinity matrices

The above gels are suitable for conventional, low-pressure chromatography systems. They provide a high-performance analytical matrix but flow-rates are relatively low. Protein affinity matrices, based on bonded microparticulate silica gels are suitable for high-pressure chromatography (23). Experience of these supports with lectins is very limited but they will provide methods for rapid sample application (and desorption) providing that the affinity between the lectin and the applied glycopeptides is high, so that binding occurs rapidly. Pre-packed affinity-matrix columns, ready for protein coupling are efficient but very expensive and they can be used only for a single lectin. Because of this, and because of constraints on sample application rates that may be imposed by kinetic parameters of lectin binding, in the opinion of the present author high-pressure techniques are unlikely to supercede low-pressure systems in the near future.

A recent paper has described the use of a high-pressure chromatography system using the two *Ricinus communis* agglutinin lectins (RCA-I and RCA-II) coupled to an activated diol silica support (23a). The authors describe fast, high-resolution separation of N-glycanase released oligosaccharides with specificities basically similar to those identified using lectin−agarose or solution-binding methods but some subtle differences in binding properties, determined by terminal sugar residues and branching patterns, were revealed between these two homologous lectins.

3.6 Regulation of degree of coupling

Some lectins may prove to be particularly sensitive to multi-point attachment to an activated matrix. In such cases, matrices with spacer arms could be tried (see above) or the number of coupling groups on CNBr-activated Sepharose can be reduced before addition of the lectin. This can be achieved by controlled hydrolysis of the active residues. According to the manufacturers (Pharmacia), 50% of coupling activity of CNBr−Sepharose is lost following incubation of the gel at pH 8.3 (0.1 M NaHCO$_3$) for 4 h. An alternative method of reducing the number of attachment points with these gels is coupling in the pH range 7.0−7.4. The active groups of the gel interact only with non-ionized free amino groups on the protein, but at neutral pH these are predominantly in the ionized form and unavailable for coupling.

As previously discussed it is generally desirable to reduce the degree of coupling when using the endolectins.

3.7 Column preparation and application of sample

The basic chromatographic procedures are relatively straightforward.

3.7.1 Column preparation

Small bed volumes are suitable for most purposes. Exceptions would be affinity gels prepared with particularly low lectin concentrations or studies on oligosaccharides which are only retarded on an affinity column, rather than forming a stable complex (see Section 3.8). In the author's laboratory columns of bed size 8 cm × 0.65 cm are routinely used at room temperature. These are suitable for semi-preparative and analytical work. An equilibration buffer of general applicability is 10 mM Tris−HCl, pH 7.4, containing 0.15 M NaCl. Gel beds are poured from a de-gassed affinity gel slurry containing 1 vol. of settled gel to 2 vols of buffer. Four bed volumes of buffer are passed through the column under a hydrostatic head of 15 cm. The column can then be used immediately or stored at 4°C in 0.05% w/v sodium azide.

3.7.2 *Sample application*

Glycopeptide or glycoprotein samples are applied to the column in equilibration buffer with the maximum sample volume at 30% of the column volume. (Larger sample volumes may be used only if high-affinity carbohydrates are being investigated as these will rapidly form complexes with the lectin.) When the sample has entered the column, flow is stopped for 15 min to ensure that binding proceeds to completion. The column is then connected to a peristaltic pump and eluted at 10 ml/h with 15 ml equilibration buffer to remove non-binding material and then with 15 ml of the same buffer containing 1 M NaCl and 1% v/v Triton X-100 to desorb any components that may be associated with the matrix through non-specific ionic or hydrophobic interactions. The non-binding fraction should be re-applied to a freshly-equilibrated column to check that lack of reactivity was not due to column saturation.

3.8 **Desorption of bound carbohydrates**

This is achieved by elution of the column with the appropriate concentration of competing sugars made up in the equilibration buffer. Step-elution or gradient-elution may be used. For some lectins, such as Con A and LCA the relationship between the structure of a bound glycopeptide or oligosaccharide and the sugar concentration required for desorption from a lectin affinity gel has been carefully studied (12,24,25). Therefore step-elution with α-methyl mannoside (α-MM) can yield carbohydrates of composition and sequence that can be predicted with reasonable accuracy. Other lectins may give information only about a 'specific' sugar (e.g. *Griffonia simplicifolia* Ia and Ib) but with little indication of sugar sequences in the vicinity of that sugar. Gradient-elution is more commonly used when the primary sugar-binding specificity of a lectin is known but when the influence on lectin-binding activity of other sugar components in a complementary oligosaccharide are less well established. Gradients are always preferable for identifying minimal structural differences between closely related oligosaccharides.

Before pumping solutions of desorbing sugars onto the column, surplus buffer should be removed from the top of the gel and replaced with about 0.5 ml of the eluting sugar solution. Columns of LCA and Con A are step-eluted at 10 ml/h with 20 ml of 5 mM α-MM and to release low-affinity glycans, composed mainly of complex N-linked oligosaccharides and then with 0.2 M α-MM to desorb high-affinity structures including, for Con A, the 'oligomannosyl' or 'high-mannose' oligosaccharides (see *Table 1*). It is quite common for elution to be carried out in a single step with $0.1-0.2$ M of competing sugar. This approach will work most successfully with the class Ia, or obligate exolectins, though the presence of microheterogeneity in the bound sample will not be detected.

When sugar gradient-elution is used for saccharide desorption the range and steepness of the gradient needs to be adapted to the experimental conditions. A useful linear gradient for preliminary studies would extend from $0-0.2$ M sugar in equilibration buffer, in a total volume of 100 ml. One ml fractions should be collected at a flow-rate of 10 ml/h.

3.9 Serial lectin affinity chromatography

This is an important refinement in technique. It provides a relatively simple and efficient means to exploit the diverse carbohydrate-binding properties of lectins for the analysis of complex mixtures of glycopeptides.

In this technique two or more affinity columns are connected in series. The sample is pumped onto the first affinity column, lectin binding carbohydrates are retained but the non-binding fraction passes onto the next affinity column containing a lectin with different sugar-binding properties from the first lectin in the series. The non-binding fraction from the second column can pass directly onto a third column and so on. A dye marker, such as phenol red, is added to the sample to monitor the flow of material through the column series. When the dye marker has cleared the final column, the columns are disconnected and eluted separately with competing sugars as described above. The procedure is illustrated in *Figure 2a* (p. 214) in which the LCA, Con A and WGA affinity columns are connected in series, in the stated order. Samples are then pumped through the column series and then each column is separately step-eluted with appropriate competing sugars. This particular column series is very useful because LCA and Con A bind the complex, high-mannose and hybrid N-glycans leaving the polylactosamine-substituted N-glycans and glycopeptides with 'clusters' of sialylated O-linked oligosaccharides to bind with the WGA column (see *Table 1*, p. 210, for structural details). In a further refinement of this technique, the Con A column can be used to distinguish between the fucosylated bi-antennate and tri-antennate oligo-saccharides that bind to LCA (*Table 1*). The LCA-binding fraction is separated from the eluting sugar α-MM, by gel filtration or by dialysis using low-molecular-weight cut-off (<1000) dialysis membrane. The sample is then applied to the Con A column which only retains the bi-antennate structures, the tri-antennate components pass through unretarded (see *Table 1*).

We have used the affinity columns illustrated in *Figure 2a* to examine cell surface glycopeptides isolated from a myeloid cell line (10,26). The glycopeptides were metabolically labelled with either [^{3}H]glucosamine or [^{3}H]mannose. As noted in Section 3.2, the [^{3}H]mannose label is highly-selective for N-linked glycans. The results are given in *Figure 3a* and *b*. The affinity columns bound about 80% of the applied cell surface glycopeptides representing a very efficient 'capture' of radiolabelled material. The profiles for the two isotopes are different due to the structural differences in the lectin binding glycans. For example, the high concentration of [^{3}H]mannose compared to [^{3}H]glucosamine in the strong affinity Con A-binding fraction arises because this fraction is enriched in the oligomannosyl or high-mannose N-glycans. By contrast, the WGA-binding fractions label more efficiently with [^{3}H]glucosamine because these materials contain polylactosamine sequences with multiple Gal−GlcNAc repeats attached to a trimannan core (*Table 1*). By reference to the structures described in *Table 1*, a close approximation to the oligosaccharides present in the LCA- and Con A-binding fractions in *Figure 3* (p. 215) can be obtained. The WGA-binding fractions should be further analysed by gel filtration on Sephadex G-100 to separate the high-molecular-weight polylactosamine and clustered O-linked glycans (structures 'h' and 'i', *Table 1*) from the smaller bisected oligosaccharides (structures 'f' and 'g', *Table 1*, ref. 10). The clustered O-linked glycans can be easily identified by their susceptibility to alkaline − borohydride treatment. Thus by using lectin affinity chromatography, together with some simple gel filtration procedures, a realistic picture can be obtained of the carbohydrate structures in a complex mixture of cell surface glycopeptides. These glycopeptides would have constituted an analytical nightmare without the use of affinity chromatography involving lectins with relevant sugar-binding specificities.

When lectins have different monosaccharide inhibitors, a variation on the above scheme for serial affinity chromatography can be used. This is illustrated in *Figure 2b*. LCA and RCA columns are connected in series and a sample is pumped through both columns. The columns are disconnected and the RCA column step-eluted with galactose. The columns are then reconnected and the LCA column eluted with α-MM. The LCA-binding fraction flows directly onto the RCA column, which will retain those oligosaccharides in the LCA-binding fraction with exposed galactose units. These can then be eluted with a second application of a galactose solution.

3.10 'Retarded' mode of carbohydrate binding

The foregoing lectins are all capable of retaining complementary oligosaccharides (K_d $< 10^{-6}$). Some lectins may show weak affinity for isolated carbohydrates which may only be retarded on lectin affinity columns rather than retained on the columns as in the previous examples. The retarded mode of carbohydrate binding is illustrated by the phytohaemagglutinins, E-PHA and L-PHA, which are both heterotypic endolectins (class IIb) that recognize complex N-linked glycan chains (*Table 1*). These lectins should be used with relatively large affinity columns (minimum size 1 cm × 20 cm) so that retardation by the gel can be accurately determined. On large pore lectin affinity gels, such as CNBr-activated Sepharose, a small glycopeptide can be described as retarded if it elutes after the column V_t. This value can be obtained by determining the volume of buffer required to elute a neutral monosaccharide such as glucose or mannose. It should be noted, however, that large glycopeptides or intact glycoproteins may contain two complementary oligosaccharides for E- or L-PHA. The bivalency of these molecules will enhance considerably their lectin affinities and they will probably be retained on lectin affinity columns under conditions in which the single oligosaccharide would only be retarded. Elution of the bound glycopeptides can be achieved by using a low pH buffer, such as 0.1 M glycine−HCl, pH 2.8. Alternatively borate buffers, which form complexes with *cis*-diols, may prove to be effective eluants.

Lectins which are capable of retaining high affinity oligosaccharides can also be used in the retarded mode to detect weak affinity interactions. An interesting example is the use of an RCA−agarose gel to fractionate a mixture of glycopeptides isolated from IgG (12). A 1 cm × 50 cm column was used in this study and applied samples were desorbed by prolonged elution with buffer (0.15 M NaCl in 20 mM Tris−acetate, pH 7.4). Three groups of non-sialylated, bi-antennate glycopeptides were identified. An unretarded fraction had two GlcNAc terminals,

$$
\begin{array}{l}
\text{GlcNAc—Man} \\
\qquad\qquad\quad \searrow \\
\qquad\qquad\qquad\quad \text{Man—GlcNAc—GlcNAc—Asn} \\
\qquad\qquad\quad \nearrow \qquad\qquad\qquad\qquad | \\
\text{GlcNAc—Man} \\
\qquad\qquad\qquad\qquad\qquad\qquad\qquad\qquad \text{Fuc}
\end{array}
$$

whereas the most retarded fraction, which required prolonged elution (3−4 column volumes of buffer) contained two end-chain galactose units,

$$
\begin{array}{c}
\text{Gal—GlcNAc—Man} \\
\hspace{2.5cm}\searrow \\
\hspace{3.5cm}\text{Man—GlcNAc—GlcNAc—Asn} \\
\hspace{2.5cm}\nearrow \hspace{2.5cm}| \\
\text{Gal—GlcNAc—Man} \hspace{2.5cm}\text{Fuc}
\end{array}
$$

A third fraction, which eluted before the above structure had only one end-chain galactose. The sialylated derivative of the above glycan is structure e, *Table 1*.

4. CHOICE OF LECTIN

The carbohydrate components of mammalian glycoproteins are very amenable to analysis by lectins. Almost without exception serum glycoproteins contain high-mannose and complex N-linked oligosaccharides and the branching patterns in the mannan core can be explored using the lectins described in *Table 1*. Clues about the microheterogeneity in the antennae may be gained from the binding strength of the oligosaccharides for a particular lectin. For example high affinity interactions with LCA or Con A, i.e. oligosaccharides released by 0.2 M, α-MM (see e.g. *Figure 3a* and *b*) are indicative of a reduction or absence of terminal sialic acid residues. The presence of galactose or GlcNAc terminals in such oligosaccharides can then be confirmed by measuring their reactivity with one of the exolectins described in *Table 2*. Cell membrane glycoproteins contain similar types of carbohydrate chain to those found in serum glyco-proteins but in addition they often contain polylactosamine-type glycans and thus WGA is a valuable analytical tool in studies of these molecules (see *Figure 3*). In mucous glycoproteins, where over 80% of the polymer may be carbohydrate (27), the oligo-saccharides are O-linked, not N-linked, and lectins such as Con A, LCA and E- or L-PHA are inappropriate. However, polylactosamines and sialic acid residues are common in mucins and the glycans can be profitably studied using WGA or *Limax flavas* lectin (*Table 2*). Mucins may also contain the same oligosaccharides that specify the blood group antigens on red cells. The group 'O' determinant is specified by the terminal trisaccharide sequence:

$$
\text{Fuc} \xrightarrow{\alpha 1,2} \text{Gal} \xrightarrow{\beta 1,4} \text{GlcNAc} \xrightarrow{\beta 1} \text{R}
$$

(where R = additional sugar residues) and this can be detected with the *Ulex europeus* lectins (2).

The blood group A and B determinants are tetrasaccharides in which GalNAc(A) or Gal(B) are attached to the above trisaccharide:

$$
\text{Gal (or GalNAc)} \xrightarrow{\alpha 1,4} \text{Gal} \xrightarrow{\beta 1,4} \text{GlcNAc} \xrightarrow{\beta 1} \text{R}
$$
$$
\hspace{2.5cm}\Big|\, \alpha 1,2
$$
$$
\hspace{2.5cm}\text{Fuc}
$$

A and B determinants may be detected by the *Griffonia simplicifolia* exolectins and one of several of the GalNAc-specific exolectins described in *Table 2* can be used to detect the A-determinant. In ref. 2 a listing may be found of highly specific endolectins that interact with blood group active carbohydrates. Another useful lectin for studying mucins is peanut agglutinin (PNA). This lectin binds to the disaccharide sequence

$$\text{Gal} \xrightarrow{\beta 1,3} \text{GalNAc}$$

which occurs in the protein-linkage region of mucins, especially in mucins produced by malignant mucosal epithelial cells (2).

In conclusion, it may be reasonably assumed that if a protein contains covalently-linked carbohydrate, then an affinity chromatographic column based on a lectin or panel of lectins, can be selected for detailed evaluation of the monosaccharide composition, sequence and branching patterns. Such 'lectin' affinity chromatography is a rapid, high-resolution non-destructive analytical procedure that can be applied to trace quantities of material. New lectins are constantly being discovered, so we can safely anticipate the availability of progressively more refined lectin analytical reagents.

5. REFERENCES

1. Goldstein,I.J. and Hayes,C.E. (1978) *Adv. Carbohydrate Chem. Biochem.*, **35**, 128.
2. Gallagher,J.T. (1984) *Biosci. Rep.*, **4**, 621.
3. Lis,H. and Sharon,N. (1986) *Annu. Rev. Biochem.*, **55**, 35.
4. Barondes,S.H. (1981) *Am. Rev. Biochem.*, **50**, 207.
4a. Osawa,T. and Tsuji,T. (1987) *Annu. Rev. Biochem.*, **56**, 21.
5. Nagata,Y. and Burger,M.M. (1974) *J. Biol. Chem.*, **249**, 3116.
6. Cummings,R.D. and Kornfeld,S. (1982) *J. Biol. Chem.*, **257**, 11235.
7. Tarentino,A.L., Gomez,C.M. and Plummer,T.H. (1985) *Biochemistry*, **24**, 4665.
8. Takasaki,S., Mizuochi,T. and Kobata,A. (1982) In *Methods in Enzymology*. Ginsburg,V. (ed.), Academic Press Inc., New York and London, vol. 83, p. 262.
9. Montreuil,J. (1980) *Adv. Carbohydrate Chem. Biochem.*, **37**, 157.
10. Gallagher,J.T., Morris,A. and Dexter,T.M. (1985) *Biochem. J.*, **231**, 115.
10a. Green,E.D. and Baenziger,J.U. (1987) *J. Biol. Chem.*, **262**, 12018.
11. Baenziger,J. and Fiete,D. (1979) *J. Biol. Chem.*, **254**, 803.
12. Kornfeld,K., Reitman,M.L. and Kornfeld,R. (1981) *J. Biol. Chem.*, **256**, 6633.
13. Debray,H., Decout,D., Strecker,G., Spik,G. and Montreuil,J. (1981) *Eur. J. Biochem.*, **117**, 41.
14. Krusius,T., Finne,J. and Raavala,H. (1976) *FEBS Lett.*, **71**, 117.
15. Cummings,R.D. and Kornfeld,S. (1982) *J. Biol. Chem.*, **257**, 11230.
16. Hammarstrom,S., Hammarstrom,M.L., Sundblad,G. and Lonngren,J. (1982) *Proc. Natl. Acad. Sci. USA*, **79**, 1611.
17. Yamashita,K., Hitoi,A. and Kobata,A. (1983) *J. Biol. Chem.*, **258**, 14753.
18. Bhavanandan,V.P. and Katlic,A.W. (1979) *J. Biol. Chem.*, **254**, 4000.
19. Schachter,H. (1978) In *The Glycoconjugates*. Horowitz,M.I. (ed.), Academic Press, New York, vol. 2, p. 88.
20. Beeley,J.G. (1985) In *Glycoprotein and Proteoglycan Techniques*. Elsevier, Amsterdam.
21. Davey,M.W., Sulkowski,E. and Carter,W.A. (1976) *Biochemistry*, **17**, 704.
22. Lotan,R., Beattie,G., Hubbell,I.W. and Nicolson,G.L. (1977) *Biochemistry*, **16**, 1787.
23. Renauer,D., Oesch,F., Kinkel,J., Unger,K.K. and Wieser,R.J. (1985) *Analyt. Biochem.*, **151**, 424.
23a. Green,E.D., Brodbeck,R.M. and Baenziger,J.U. (1987) *J. Biol. Chem.*, **262**, 12030.
24. Yamamoto,K., Tsaji,T. and Osawa,T. (1982) *Carbohydrate Res.*, **110**, 283.
25. Narasimhan,S., Wilson,J.R., Martin,E. and Schachter,H. (1979) *Can. J. Biochem.*, **59**, 83.
26. Morris,A., Gallagher,J.T. and Dexter,T.M. (1986) *Biomed. Chromatogr.*, **1**, 41.

Manufacturers and suppliers of HPLC columns and column support materials[a], and of HPLC instruments[b]

R.W.A.OLIVER

It should be noted that many companies have offices and manufacturing plants in different countries. Where possible the name and addresses of the UK and the USA locations are given, as well as that of the parent company if not in either of these countries.

Affinity Purification Ltd[a], The Innovation Centre, The Science Park, Cambridge CB4 4GF, UK

Alltech Assoc. Inc.[a], 2051 Waukegan Road, Deerfield, IL 60015, USA

Altex Scientific (Division of Beckman)[a,b], 1780 Fourth Street, Berkeley, CA 94710, USA

Amicon Ltd[a,b] (Division of W.R.Grace), Upper Mill, Stonehouse, Gloucester GL10 2BJ, UK

Anachem Ltd[a], Charles St, Luton, Bedfordshire LU2 0EB, UK

Analabs Inc.[a] (A unit of Foxboro Analytical), 80 Republic Drive, North Haven, CT 06473, USA

Anspec Co. Inc.[b], 50 Enterprise Drive, PO Box 7730, Ann Arbor, MI 48107, USA

Applied Biosystems Ltd[b] (formerly Kratos Analytical Systems), Birchwood Science Park North, Warrington, Cheshire WA3 7PB, UK / 850 Lincoln Centre Dr., Foster City, CA 94404, USA

Applied Chromatography Inc., PO Box 1008, State College, PA 16804, USA

Applied Chromatography Systems Ltd[a,b], The Arsenal, Heapy Street, Macclesfield, Cheshire SK11 7JB, UK

Applied Science Laboratories Inc.[a], PO Box 440, State College, PA 16801, USA

Autochrome Inc.[b], PO Box 207, Milford, MA 01757, USA

J.T.Baker Chemicals[a], PO Box 1, 7400AA, Deventer, Holland

J.T.Baker Chemical Co.[a], 222 Red School Lane, Phillipsburgh, NJ 08865, USA

Beckman Instruments Altex Div.[a,b], PO Box 5105, 2350 Camino Ramon, San Ramon, CA 94583, USA

Beckman Instruments International S.A., 17 Rue des Pierres-du-Niton, PO Box 76, CH-1207, Geneva 6, Switzerland

Beckman Instruments UK Ltd[a,b], Progress Road, Sands Industrial Estate, High Wycombe, Buckinghamshire HP12 4JL, UK

Bioanalytical Systems Inc.[a,b], 1205 Kent Ave, W.Lafayette, IN 47906, USA

Bio-Rad Laboratories Ltd[a,b], Caxton Way, Watford Business Park, Watford, Hertfordshire WD1 8XG, UK / 2200 Wright Avenue, Richmond, CA 94804, USA

Brownlee Laboratories Inc.[a], 2045 Martin Avenue, Santa Clara, CA 95050, USA

Bruker Analytische Messtechnik GmbH.[b], Silberstreifen, D-7512 Rheinstetten 4, FRG

Bruker Instruments Inc.[b], Manning Park, Billerica, MA 01821, USA

Bruker Spectrospin Ltd[b], Unit 3, 209 Torrington Avenue, Coventry, West Midlands CV4 9HN, UK

Carlo Erba Instruments (UK)[b], c/o MSE Scientific Instruments, Sussex Manor Park, Gatwick Road, Crawley, West Sussex, UK

Carlo Erba Strumentazione SpA.[b], Strada Rivoltana, 1-20090 Rodano, Milan, Italy

Cecil Instruments Ltd[b], Milton Industrial Estate, Cambridge, CB4 4AZ, UK

Chromanetics Corp.[a], 9544 Belair Road, Baltimore, MD 12236, USA

Chromatec Inc.[a], 30 Main Street, Ashland, MA 01721, USA

Chromatix Inc.[a], 560 Oakmead Parkway, Sunnyvale, CA 90607, USA

Chrompack[a], Unit 4, Indescon Court, Millharbour, London E14 9TN, UK

Chrompack International BV[a], PO Box 8033, 4330 Middleburgh, The Netherlands

Chrompack USA Inc.[a], 14802 Janine Dr., Whittier, CA 90607, USA

Dionex Corp.[b], 1228 Titan Way, Sunnyvale, CA 94086, USA

Dionex (UK) Ltd[b], Eelmoor Road, Farnborough, Hampshire GU14 7QN, UK

Du Pont de Nemours[a,b], Chandler Buildings, Wilmington, DE 19898, USA

Du Pont de Nemours (Deutschland) GmbH., Biotechnology Systems Div.[a,b], Postfach 401240, D-6072 Dreieich, FRG

Du Pont (UK) Ltd[a,b], Wedgwood Way, Stevenage, Hertfordshire SG1 4QN, UK

Durrum Chemical Co.[a,b], 3950 Fabian Way, Palo Alto, CA 94303, USA

Dyno Particles A.S.[a], PO Box 160, N-2001, Lillestrøm, Norway

Electro-Nucleonics Inc.[a], 350 Passaic Ave, Fairfield, NJ 07006, USA

Electro-Nucleonics Intp. Ltd[a], Adriaan Van Bergenstraat 202−208, SW Breda NL 4811, The Netherlands

EM Science/Hitachi[a,b], 111 Woodcrest Road, Cherry Hill, NJ 08034, USA

ESA Inc.[b], 45 Wiggins Ave., Bedford, MA 01730, USA

Foxboro Analytical Co.[b], PO Box 449, S.Norwalk, CT 06856, USA

Foxboro Great Britain Ltd[b], Wiggle House, Redhill, Surrey RH1 2HL, UK

Gilson Medical Electronics Inc.[b], c/o Anachem Ltd, 20 Charles Street, Luton, Bedfordshire, UK / 3000 W.Beltine Hwy, PO Box 27, Middleton, WI 53562, USA

Glenco Scientific Inc.[a,b], 2802 White Oak, Houston, TX 77007, USA

Gow-Mac Instruments Co.[a,b], PO Box 32, Bound Brook, NJ 08805, USA

W.R.Grace & Co. (Amicon Div.)[a,b], 24 Cherry Hill Drive, Danvers, MA 01923, USA

Hamilton Co.[a], PO Box 10030, Reno, NV 89510, USA

Haskel Energy Systems Ltd[b], North Hylton Road, Sunderland SR5 3JD, UK

Haskel Inc.[b], 100-88 E.Graham Place, Burbank, CA 91502, USA

Hewlett Packard Co. Analytical Group[b], Mailstop 20B3, PO Box 10301, Palo Alto, CA 94303, USA

Hewlett-Packard Ltd[b], Miller House, The Ring, Bracknell, Berkshire RG12 1XN, UK

Hitachi Scientific Instruments[a,b], 460 E.Middlesfield Road, Mountain View, CA 94043, USA

HPLC Technology Ltd[a,b], Wellington House, Waterloo Street West, Macclesfield, Cheshire SK11 6PJ, UK

IBM Instruments Inc.[a,b], PO Box 3332, Danbury, CT 06813, USA

Instrumentation Specialists Inc. (ISCO)[a,b], 4700 Superior Street, Lincoln, NE 68506, USA

Instrumentos Scientificos CG[b], Avenue Vereador, José Dinez 2421, 04603 Saõ Paolo, Brazil

Interaction Chemicals Inc.[a], 1615 Plymouth Street, Mountain View, CA 94043, USA

Jasco[b], c/o Ciba Corning Diagnostics, Halstead, Essex CO9 2DX, UK

Jasco International[b], 218 Bay Street, Easton, MD 21601, USA

Jobin−Yvon[b], 18 Rue de Canal, 91160 Longjumeau, France

Johns-Manville[a], Ken-Caryl Ranch, Denver, CO 80217, USA

Jones Chromatography Inc.[a], PO Box 132018, Columbus, OH 43213, USA

Jones Chromatography Ltd[a], Tir-y-Berth, Ind. Estate, Mid Glamorgan, UK

Kipp & Zonen[b], 390 Central Ave, Bohemia, NY 11716, USA

Kipp & Zonen Delft BV[b], Mercuriusweg, 1, NL-2624 BC Delft, The Netherlands

Dr Herbert Knauer, Wissenschaftiche Gerate KG[b], Heuchelheimer Strasse 9, bad Homberg v.d.H. D-6380, FRG

KNAUER/Sonntek[b], PO Box 8589, Woodcliffe Lake, NJ 07675, USA

Koken Co. Ltd[a], Toshima-ku, Tokyo, Japan

Kontron AG Instruments Ltd[a,b], Blackmoor Lane, Croxley Centre, Watford, Herts WD− 8XQ, UK

Kontron Instruments[a,b], 9 Plymouth Street, Everett, MA 02149, USA

Laboratory Data Control (LDC/Milton Roy)[b], Interstate Industrial Park, PO Box 10235, Riviera Beach, FL 33404, USA

LDC/Milton Roy UK Ltd[b], Milton Roy House, Diamond Way, Stone Business Park, Stone, Staffordshire ST15 0HH, UK

Laboratory Instrument Works[a], Prague, Czechoslovakia

Machery−Nagel GmbH.[a], Neumann-Neander-Str., PO Box 307, D-5160 Dueren, FRG

Merck/Hitachi[a,b], Frankfurter Str., 250 Postfach 4119, D-6100, Darmstadt, FRG

Micromeritics Instruments Corp.[b], 1 Micromeritics Drive, Norcross, GA 30093-1877, USA

Mitsubishi Chemical Industries Ltd[a], Marunouchi 2-chome, Chiyoda-ku, Tokyo 100, Japan

Mitsui Toatsu Chemicals Inc.[a], Chiyoda-ku, Tokyo, Japan

Molecular Separations[b], PO Drawer E, Champion, PA 15622, USA

Nicolat Analytical Instruments[b], 5225-1 Verona Road, Madison, WI 53711, USA

Nicolet Instruments Ltd[b], Budbrooke Road, Warwick, CV34 5XH, UK

Packard-Becker BV[b], Postbus 519, Delft, The Netherlands

Pentax Handelsgellschaft GmbH.[a], 2000 Hamburg 54, FRG

Perkin-Elmer Corporation, Instruments Div.[a,b], Main Avenue, Norwalk, CT 06856, USA

Perkin-Elmer Ltd[a,b], Post Office Lane, Beaconsfield, Buckinghamshire HP9 1QA, UK

Perstorp Biolytica AB[a], S-223 70, Lund, Sweden

Pharmacia Ltd/LKB[a,b], Pharmacia House, Midsummer Boulevard, Milton Keynes, MK9 3HP, UK

Pharmacia LKB Biotechnology Inc.[a,b], 800 Centennial Ave, Poscataway, NJ 08854, USA

Phase Separations Ltd[a], Unit 19, Deeside Industrial Park, Queensbury, Clwyd CH5 2NU, UK

Phase Separations Inc.[a], 140 Water Street, Norwalk, CT 06854, USA

Phase Separations BV[a], Staringlann 21, 2741 GC Waddinxveen, The Netherlands

Phillips Analytical Div.[a,b], Bldg. HKF, NL-5600 MD Eindhoven, The Netherlands

Phillips Analytical Pye Unicam Ltd[a,b], York Street, Cambridge, CB1 2PX, UK

Pierce Chemical Co.[a], PO Box 117, Rockford, IL 61105, USA

Pierce Eurochemie BV[a], PO Box 1512, NL 3260 BA Oud-Beijerland, The Netherlands

Pierce UK Ltd[a], 36 Clifton Road, Cambridge CB1 4ZR, UK

Poly LC[a], 9052 Bellwart Way, Columbia, MD, USA

Polymer Laboratories Inc.[a], 160 Old Farm Road, Technical Center, Amherst Fields Research Park, Amherst, MA 01002, USA

Polymer Laboratories Ltd[a], Essex Road, Church Stretton, Shropshire SY6 6AX, UK

Rainin Instrument Co. Inc.[b], Mack Road, Woburn, MA 01801, USA

Regis Chemical Co.[a], 8210 Austin, Morton Grove, IL 60053, USA

Rheodyne Inc.[b], PO Box 996, Cotati, CA 94928, USA

RSL[a], Begoniastraat 5, B-9731, Nazareth (Eke), Belgium

Scientific Systems Inc.[a], 1120 West College Ave, State College, PA 16801, USA

Separations Technology Inc.[a], PO Box 63, 2 Colombia Street, Wakefield, RI 02879, USA

Serva Feinbiochemica GmbH.[a], D-6900, Heidelburg 1, FRG

Shandon Southern Inc.[a], 515 Broad Street, Sewickley, PA 15143, USA

Shandon Southern Products Ltd[a], 93 Chadwick Road, Runcorn, Cheshire WA7 1PR, UK

Shimadzu Scientific Instruments[b], c/o Dyson Instruments Ltd, Hetton Lyons Industrial Estate, Houghton-le-Spring, Tyne and Wear DH5 0RH, UK

Shimadzu Scientific Instruments Inc.[b], 7102 Riverwood Road, Columbia, MD 21046, USA

Showa Denko[a,b], 34 Shipa Miyamoto-cho, Minato-ku, Tokyo J105, Japan

Showa Denko America Inc.[a,b], 280 Park Ave, New York, NY 10017, USA

Siemens AG[b], Abt. Rontgenanalyse, E. 689, Postfach 211080 D-7500 Karlsruhe 21, FRG

Siemens Ltd[b], Siemens House, Eaton Bank, Congleton, Cheshire CW12 1PH, UK

SOTA Chromatography Inc.[a], PO Box 693, Crompond, NY, USA

Spectra-Physics[b], 2905 Stender Way, Santa Clara, CA 95051, USA

Spectra-Physics Ltd[b], 17 Brick Knoll Park, St Albans, Herts AL1 5UF, UK

Supelchem UK[a], London Road, Sawbridgworth, Herts. CM21 9JH, UK

Supelco Inc.[a], Supelco Park, Bellefonte, PA 16823, USA

Synchon Inc.[a], PO Box 310, Lafayette, IN 47902, USA

Technicol Ltd[a], Brook Street, Higher Hillgate, Stockport, Cheshire SK1 3HS, UK

The Separations Group (Vydac)[a], PO Box 867, 17434 Mojave Street, Hesperia, CA 92345, USA

Toya Soda Manufacturing Co. Ltd[a], 1-14-15 Akasaka, Tokyo 107, Japan

Tracor Atlas Ltd[a,b], Unit 1, 5 Millbrook Industrial Estate, Crowborough, East Sussex TN6 3DZ, UK

Tracor Instruments (Division of Tracor Inc.)[a,b], 6500 Tracor Lane, Austin, TX 78721, USA

Unimetrics Corporation[b], 501 Earl Road, Shorewood, IL 60436, USA

Universal Scientific Inc.[b], Suite 101, 2070 Peachtree Industrial Court, Atlanta, GA 30341, USA

Varex Corporation[b], 12221 Parklawn Drive, Rockville, MD 20852, USA

Varian Associates Inc.[a,b], 220 Humboldt Court, Sunnyvale, CA 94089, USA

Varian Associates Ltd[a,b], European Marketing Headquarters, 28 Manor Road, Walton-on-Thames, Surrey KT12 2QF, UK

Waters Chromatography (Division of Millipore)[a,b], 11–15 Peterborough Road, Harrow, Middlesex, HA1 2YH, UK

Waters Millipore, Chromatography Div.[a,b], 34 Maple Street, Milford, MA 10757, USA

Wescan Instruments[b], 2051 Waukegan Road, Deerfield, IL 60015, USA

Whatman Chemical Separations Inc.[a], 9 Bridewell Place, Clifton, NJ 07014, USA

Whatman Ltd[a], Springfield Mill, Maidstone, Kent ME14 2LE, UK

Woelm Pharmaceuticals[a], c/o Universal Scientific Inc., PO Box 80402, 1970 Peachtree Ct, Atlanta, GA 30341, USA / PO Box 840, D-3440 Eschwege, FRG

YMC Inc., c/o Hichrom Ltd, 6 Chiltern Enterprise Centre, Station Road, Theale, Reading RG7 4AA, UK / 51 Gibraltar Drive, Morris Plains, NJ 07950, USA

INDEX